THE
FORENSIC
CASEBOOK

THE FORENSIC CASEBOOK

THE SCIENCE OF CRIME SCENE INVESTIGATION

N.E. GENGE

EBURY
PRESS

First published in the US by Ballantine

This edition published in Great Britain in 2004

5 7 9 10 8 6 4

Ebury Press, an imprint of Ebury Press.
Random House, 20 Vauxhall Bridge Road, London SW1V 2SA

Random House Australia (Pty) Limited
20 Alfred Street, Milsons Point, Sydney, New South Wales 2061, Australia

Random House New Zealand Limited
18 Poland Road, Glenfield, Auckland 10, New Zealand

Random House (Pty) Limited
Isle of Houghton, Corner of Boundary Road & Carse O'Gowrie, Houghton 2198, South Africa

The Random House Group Limited Reg. No. 954009

www.randomhouse.co.uk

A CIP catalogue record for this book is available from the British Library

ISBN 9780091897284 (from Jan. 2007)

ISBN 0091897289

Book design by Joseph Rutt

Printed and bound in Great Britain by William Clowes Ltd, Beccles, Suffolk

For my mother, Mamie Sainsbury—
one of the many who have helped victims of violence

CONTENTS

ACKNOWLEDGMENTS

Throughout this book, you will find the names of the dozens of forensic specialists, law enforcement officers, emergency personnel, and criminalists who have generously shared their knowledge and donated their experiences. If I have any regret about this project, it is only that I couldn't include more of their intriguing stories. It's rare indeed for any group of people to share their mistakes as freely as their successes and, as I look back at the finished manuscript, I can't help but be struck, once again, by their integrity, both personal and professional.

Special thanks to the Royal Newfoundland Constabulary and Constable Doug Miller for photographic access and assistance. Additionally, I'm indebted to Ling Lucas, who remains as terrific a person as she is an agent, and to Ku and the staff at Ballantine who've made the completion of this book such a pleasure.

As always, none of this would be possible without Peter and Michael, two of the finest people I know.

THE
FORENSIC
CASEBOOK

ONE

THE SCENE OF THE CRIME

The Scene of the Crime Versus the Crime Scene

For novelists like Agatha Christie, the scene of the crime was obvious—a blood-splattered drawing room or the gaping wall safe in a society grand dame's bedroom. Modern law enforcement officers, criminalists, forensic scientists, and even the often maligned private investigators, however, have quite different definitions.

The "scene of the crime" might be a room, an entire building, a wharf and surrounding harbor, the three-mile path of a disintegrating airplane, or just the keypad and receiver of a public phone booth. However, the actual "crime scene," to use the language of forensic investigation, often bears little resemblance to the physical layout visible at the scene. And, surprisingly to most, the size of the scene often has little relation to the time required to work it. For example, a phone booth regularly used by a stalker in Kingston, Australia, required eight technicians working for three full days. The first crash scene David Kellerman of the Aeronautics Investigative Unit ever worked required dozens during the rescue phase, but the actual forensic investigation—due in large part to the first responder's swift creation of a single access path for rescuers, as well as to the presence of snow (which helped in locating loose debris some distance from the

crash)—took a mere nine investigators less than two days to complete.

Of course, the physical size of the crime scene bears no relation whatsoever to the number of individuals, either suspects or witnesses, involved and, as collecting statements is as much a part of "working" the crime scene as swabbing blood stains, scene size can be quite deceptive.

A nylon bag, measuring less than a foot square and four inches thick and containing nothing more than a laptop and modem, was the only physical evidence of a conspiracy involving hundreds of people and millions of dollars in a penny-stock scam. At the other extreme, the bombing of the Alfred P. Murrah Building in Oklahoma City was ultimately attributable to only a handful of people. The number of witness statements required during these investigations, however, was virtually identical.

The tiny computer bag, every bit as much a crime scene as the rubble of the Murrah Building, presented investigators with specialized problems not found at scenes requiring miles of yellow tape. The bag was mobile, its contents were subject to destruction by the simple wave of a magnet or flip of a switch, and the computer inside was the actual property of any number of suspects. The scene in Oklahoma, while horrific in its loss of life, couldn't be slipped into a pocket and hidden away from analysts.

A "crime scene" isn't only the actual location of the crime—it is also the staging and planning areas, the paths of flight to and from the primary scene, and the paths between the primary and secondary scenes. Consequently, the total crime scene for a twenty-first-century offense might prove international and include dozens of physical locations and individuals, thousands of exhibits, and nearly as many witness statements.

As I prepare this manuscript, the investigation into the September 11 attacks on the Pentagon and World Trade Center and the crash in Pennsylvania, is just swinging into high gear. Already the FBI is reporting nearly 7,000 investigators and support personnel following up "35,000+ leads" and looking into "numerous international possibilities." Cars at several airports, apartments and hotels in places as far apart as Boston, Massachusetts, and Vero Beach, Florida, are just a few of the secondary scenes being secured. Electronic scenes also could contain vital information, but such evidence could be as difficult to collect as an impression of a shoe in the snow of a sunny field.

Scenes like the Pentagon and the World Trade Center are obviously, and thankfully, the exception. But the far-reaching scope and the need to coordinate rescue efforts with forensic investigation—all while maintaining a secure scene in which the lives of investigators and emergency staff aren't further endangered—makes these cases well-documented examples of the difficulties that law enforcement personnel face, even if it is usually on a smaller scale.

Crime happens everywhere, and determining what territory, items, and persons make up the "crime scene" isn't always easy. Constantly confronted with the dangerous or unexpected, investigators juggle a number of imperatives in their efforts to turn the "scene of the crime" into a secured "crime scene." But, in the vast majority of cases, this is the first decision to be made in a criminal investigation, and it is often made by the first responder to the scene—not a criminalist. A mistake made at this stage can end any chance of solving the crime, let alone prosecuting the offenders.

Both the film version and the book version of Anna Porter's *The Bookfair Murders* begin with the crime scene investigator's worst nightmare: a murder accomplished in the middle of 300 witnesses—300 potential suspects—all of whom are due to scatter to numerous countries outside the local detective's jurisdiction in three days if the investigation doesn't generate enough evidence to book someone. On top of that, the Frankfurt Bookfair is a four-day event, with conference rooms and booths set up in temporary quarters through which thousands of people tramp each and every day of the show.

As the collection of evidence at the crime scene includes the names and contact information—and, hopefully, statements—of everyone present, the detective on this case could have spent the entirety of those four days doing little more than arranging interviews!

Protecting the Scene: The First Responder

The first person on the scene is immediately confronted with a number of considerations: victims who may be in need of immediate attention, witnesses ready to melt away at the first opportunity, the possibility of further criminal activity, the

responsibility of preserving whatever evidence might be remaining and securing a crime scene while maintaining safe corridors for emergency personnel. This person must weigh all these needs and make immediate decisions based on the situation. And every situation is, in some way, unique.

Imagine arriving at a subway platform at 8:30 A.M. to find the victim still on the tracks, an injured passerby sitting on his briefcase, hundreds of people spilling through the scene, and a dozen points of possible egress for the perpetrator. Where, in that milling mess, does the yellow tape belong? Who else should be called in, and in what order should they be called? Is it even possible to secure this scene?

Fortunately, while each scene includes elements that could trip up an investigator, the experience of hundreds of others has provided a rough outline for what must be accomplished, as well as the order in which these tasks should be done. Faced with an overwhelming situation, the first responder can operate on autopilot while impressions sink in and decisions are made.

The Responsibilities of the First Responder

Observe and Establish the Likely Parameters of the Crime Scene

1. As the crime may be ongoing, the first responder must assume the scene is unsecured and dangerous until proven otherwise.

2. Observe the immediate scene, identifying the major physical characteristics, persons (living, deceased, injured, lucid, or confused), and paths into and away from the area. If possible, identify any individual who might have called in a crime and get their names.

3. Note the entrance and exit of any persons or vehicles during the initial scan.

4. Identify the primary scene and any possible secondary sites or paths to and from them.

5. Establish a safe path of access or wait until one can be established safely.

6. Make sensory observations, including sights, sounds, and smells.

Initiate Safety Procedures

7. Based on observations, contact any personnel needed to ensure the safety of additional responders. If there is evidence of a bomb, beware of secondary explosions designed to ensnare responding emergency personnel. If there's a possibility of biological or chemical hazards, like a natural gas leak, or the continued presence of dangerous persons, ensure that appropriate warnings are passed to everyone approaching the scene.

8. Ensure that those remaining within the scene, including victims or witnesses, are aware of any possible hazards.

Provide Emergency Care

9. Determine the status of witnesses and victims, checking for signs of life and medical needs.

10. Call for medical backup and provide first aid if required.

11. Guide medical personnel along the path that is *both* least likely to impair physical evidence at the scene and to quickly deliver them to victims.

12. Ensure medical personnel are aware of the need to preserve evidence (i.e., slugs or victims' clothing) and to preserve the crime scene as untouched as possible.

13. Find the names of facilities to which victims may be transported.

14. Get initial statements from victims who can provide them.

15. Ask medical personnel to note statements made by victims.

16. Attempt to have law enforcement personnel meet the victims at the hospital if no one is available to travel with the victims from the crime scene.

Secure the Scene

17. Beginning with those already on the scene, escort all persons to a safe and secure location from where they cannot alter the scene. To prevent discussion of the crime or the scene, it is best if persons are moved separately.

18. Make preliminary separations of witnesses, suspects,

bystanders, and begin identification, interviews, or con-
solation as appropriate.

19. Ensure that "official" bystanders (reporters, non-involved
    law enforcement personnel, etc.) do *not* gain access to
    the scene.

Physically Secure the Scene and Evidence

20. Working outward from the primary incident site, se-
    cure the primary incident site, all sites of ingress and
    egress, and any vehicles that may be associated with
    the crime, the witnesses, suspects, or victims. Use barrier
    tape, flags, flagged ropes, or other suitable materials.

21. If it appears likely that the media will become intrusive,
    arrange for visual barricades.

22. If specific apparent evidence is in danger of destruction,
    cover it or secure it in some other manner.

23. Document. Take note of everything done to date, persons
    on the scene, persons called to the scene, investigative
    animals on scene, locations of vehicles, positions of doors,
    environmental conditions, etc. Whenever possible, in-
    clude sketches, measurements, and diagrams.

24. Establish a "control center" location for allowing or deny-
    ing access to the site and a "point officer" in charge of
    passing information into and out of the site, thereby lim-
    iting the number of people crossing the line.

25. In collecting evidence, carefully consider search and seizure
    issues and contact appropriate authorities for assistance.

26. In addition to ensuring the physical security of evidence,
    it's important to try to protect the crime scene from
    atmospheric changes where possible. For instance, it's

helpful to leave AC/heating controls in place, windows open or closed, and electrical devices on or off as found. More obvious disruptions (smoking, eating, coughing, etc.) should also be avoided at the crime scene.

27. If items must be removed, first document their original locations and conditions as thoroughly as possible.

Release the Scene

28. Once appropriate authorities arrive, release the scene, documenting who has released it and to whom. Apprise new authorities of any outstanding or imminent issues.

29. Provide the control center log of personnel arriving and leaving.

30. Remain available until relieved.

31. Continue to flesh out documentation as possible.

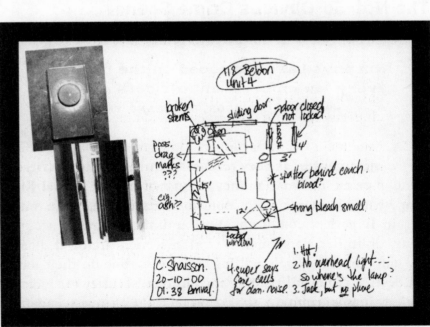

Finalize Documentation

32. Ensure that all notes are clear, all diagrams are well labeled, and, when time allows, all notes of the incident are typewritten and copied to all appropriate files and personnel. Note the environmental conditions and follow up any outstanding items.

33. Review incoming information and compare it to contemporaneous notes to find possible discrepancies.

34. Provide a narrative account of all activity occurring up until the point of release.

Now the scene of the crime is also a "crime scene" ready for criminalists, technicians, and forensic scientists—the next stage of the investigation.

Special Problems: The Not-So-Obvious Crime Scene

"Murphy's Law is tattooed on the inside of every law enforcement officer's eyelids."
—PURCELL WINDSOR, SCOTLAND YARD

Few criminalists (as they're known in the United States) or scene-of-crime officers (as they're known in the United Kingdom) would disagree with Windsor's statement. Even under seemingly perfect conditions, with unlimited time to secure a scene, unlimited personnel to work a scene, careful, dedicated law enforcement people willing to let everyone do their job to the best of their ability, and an unlimited budget for forensic work, things go wrong.

"I once spent two days searching a wall for a bullet that *had* to be there." Clayton Spencer, twelve-year veteran of one of Britain's Criminal Investigation Divisions, still shakes his head over "the one that got away." "I went back and told my super—twice—that it wasn't in that wall. She came out, brought a ballistics expert, a reconstructionist, and two detectives. We tried strings, laser alignment, you name it. We had a 'through and through' bullet wound in the victim—no way it was still in him—but the bullet just disappeared between him and the wall two meters behind him! It wasn't on the ground, down a drain, and, like I'd been saying all along, it sure wasn't in the wall. The six of us went over that wall square inch by square centimeter, nothing."

So where was it?

"We'd finally packed up, opened the scene, in fact, which was a very big deal as it was a public thoroughfare. Given up on it completely. Then, three days after the shooting, this lad comes into the station pushing his bike ahead of him. He'd been away for the weekend, come home, taken his bike out of the shed, found the tire flat and, there, right there, in the rubber tire, was our slug. Kid had ridden right through the scene—didn't even know that the chap had been shot, just zipped right on by."

Vonda McComb, working in Australia, faced a similar situation when a man attempted to shoot his wife.

"She claimed he'd shot at her, but we couldn't find evidence of a bullet anywhere in the room—a rather small bedroom, no more than eight feet across. We looked through the mattress twice, brought in a metal detector, even tried lines of sight through the only window, just in case it was open—though we were sure it wasn't at the time."

Nothing?

"Nothing. Not until her dog disappeared. She was convinced the husband had done something to him, to terrify her. We found him a couple days later. Crawled under the garage nursing his bullet wound, a wound obviously considerably older, jiving nicely with her first report. The dog was fine once the vet dug out the bullet, and we had our evidence."

Vonda can laugh about it now, but, at the time, they'd had serious concerns for the woman's safety. "Without proof that she was being threatened, we couldn't get a judge to even issue a restraining order against the husband."

Of course, dogs and bicycles aren't the only mobile objects that can travel through a crime scene. Cars do. Shopping bags do. Kids' lunch boxes do. Even people do. And any of them can be contaminating, taking something away from or adding something to the crime scene—all unnoticed even by trained investigators.

Evidence that doesn't move, however, also challenges analysts, though in different ways. For example, while much evidence at a scene is patent, or clearly visible (a broken lock or a blood splatter), much more is latent, or invisible to the eye. Prints on metal and glass aren't too difficult to see, but prints on paper or other matte surfaces can be extremely difficult to detect. Print powders—including the "dragon's blood" that fictional investigator Gil Grissom of *C.S.I.* favors for his special cases—make tough-to-spot prints visible; soft tissue X rays can even turn up prints on people, a feat once believed impossible by print technicians. But, without some expectation that evidence *might* be there, such evidence could be lost or destroyed before technicians get to it.

Some evidence truly *is* invisible, even to the most dedicated crime scene technicians. Odors at the scene, fragile evidence at the best of times, are practically impossible to collect. The English language, so rich in words, has surprisingly few that describe scents. Descriptions of odors, therefore, remain imprecise, even ambiguous.

A few examples:

"The first officer on the scene reported immediately noticing a sharp, greasy smell around the front door. It wasn't apparent to her twenty minutes later when the detective arrived."

"It was a sort of spicy smell."

"Bad, it smelled really, really bad!"

To Sniff or Not to Sniff: A Real Question

Silence of the Lambs addressed the "really, really bad" aspect of forensic investigation when Clarice Starling and her fellow officers smeared a menthol cream under their noses before tackling the preliminary postmortem of one of Buffalo Bill's victims, a body dragged out of the water.

The Bone Collector's investigator, Lincoln Rhyme, though wheelchair-bound and unable to crawl through scenes himself, demonstrates a diametrically opposed form of evidence collection. When his long-distance nose, Amelia Sachs, is repulsed at sniffing bodies in a basement known to be contaminated with free-floating asbestos, he pushes her to capture those elusive smells, even to the point of sticking her nose to the neck of a scalded body, eventually producing descriptions about as precise as one can be with our language: *She jammed her lids closed, inhaled, fought the nausea. "Mold, mustiness . . . hot water . . . cologne, dry, like gin."*

In real life, our lack of ability to differentiate scents—or to describe the ones we do sense—is occasionally compensated for by aids both mechanical and canine. Scientists studying odorology (called "the emergent science of forensic sniffing" by scent scientist Kay Dick) are creating mechanical devices—scent-busters—that transfer scent to filters for chemical analysis or use with K-9 teams. Dogs, long used to sniff drugs and bombs, are being trained to key off more unusual substances: chloroform, alcohol (a prime ingredient in colognes and perfumes), the aluminum compounds frequently found in deodorants—anything that might turn up a body, track a fleeing suspect, or make links between individuals and locations.

Phil Rendell, a medical examiner, refuses to allow menthol products into his autopsy suites or pathology labs. "We have little enough evidence available to us without deliberately ignoring yet another source, disgusting or not."

Trace evidence—hair, fibers, soil samples—provide investigators dozens of possible leads, but finding that evidence amid the bustle of the scene can make finding a needle in a haystack look easy.

Fumiko Cleal, a New York City investigator, once crawled through over a mile of air conditioning ducts—collecting evidence throughout, securing filters, generating nearly 400 exhibits—all to prove a rapist was at the scene, breaking a strong alibi and provoking a confession. "We'd found nothing at the main scene, but every time I saw the suspect, he was wearing this funky jacket. It might have been his only jacket, but it had a sort of furry collar that looked like it might shed." She'd gambled that, while the building itself was regularly cleaned, the filters would only be changed once or twice a year. "I couldn't *see* anything, even with the big light I dragged through the ducts, not until I got it back to the lab—actually, I used the big scope at the university—but Locard [the "father" of forensics] tells us there has to be something there, if we look hard enough."

Fine grains of sand were the clincher found by Ling Yu in a California home invasion and burglary. "Most people equate sand and California and ask where in California you *wouldn't* expect to find sand, right?" The two small plastic bags holding her "favorite dirt" certainly looked identical to the naked eye.

"I'd seen the initial reports from the first officers on the scene, and they'd noted an out-of-state plate, so I guess that was floating in my subconsciousness somewhere. It wasn't actually on the street outside the crime scene at that time, but this was a small community and it was noticeable to the officer. She mentioned it twice while I was in hearing distance, so . . . I guess I was ready to see something else out of place."

Still, sand?

"The rest of the house was quite tidy, even after several

officers and investigators going through, so sand, away from the front door, right at the base of a window where the family told us the burglar had stood to watch the street, was out of place, even if it didn't look different. The differences turned up in microscopy and were consistent with the home state of the license plate. We got a good conviction on that one."

Other evidence is not only tough to find and collect but also difficult to process. The spray of spittle on a telephone receiver, potentially an abundant source of DNA, is collected on faith. Though some field tests can suggest the presence of DNA, and new tests offer hope of field identification at some point in the future, most detectives swabbing that surface are indulging in the ultimate Hail Mary.

Consider this, from a National Institute of Justice flyer:

A recent FBI survey revealed that of all rapes, less than half were solved by the police, and less than 10 percent were sent to crime laboratories. And, because crime laboratories are not able to work all cases submitted, in only 6 percent of the 250,000 rape cases was the recovered DNA tested, leaving a backlog of several thousand cases awaiting processing.

Less than 6 percent. For a major crime.

Now try calculating the chances of a swab from a minor case finding its way through the process to a reputable lab, back to a department where the case is still considered active, to a detective not already swamped by other cases, to help solve a crime that only the victim is following with any interest.

With evidence measured in parts per million, the "out of sight, out of mind" adage has new meaning. Even the most diligent investigator could be forgiven for not wasting his or her time on collecting evidence that may never see the light of day in favor of more traditional exhibits that might actually yield a conviction.

Special Locations

Special locations, as opposed to unusual places where physical evidence might be found, are locations that present special considerations of forensic value.

A man found dangling from a rope and wearing his girlfriend's underwear is of no more forensic note than a man wearing anything else. We might infer something about the crime itself, but not about the evidence. Likewise, a bullet found 200 miles from the scene of a murder—in the tire of a passing car that had left the scene without anyone the wiser—is a curiosity, but it won't change the ability of a firearms expert to match the slug to the gun from which it came.

A special location needn't be rare or unusual, but it will change some aspect of the forensic study, negating the usual rules. Water, on a planet covered by oceans, lakes, and rivers, isn't rare, but it presents investigators with unique problems. Adipocere, a "soap" formed of human fat and other products degrading in water, can eventually replace all tissue. Clearly, a body thus affected presents special difficulties.

Decomposition, a fairly well-understood process in an air environment, follows entirely different pathways in a submerged body. Pressure, degree of salinity, temperature changes at various depths, and bacteria and insects particular to aquatic environments can all drastically alter the rate and course of putrefaction, making time and cause of death very difficult to determine.

And there are other examples of special locations. Extreme altitude or depth can cause chemical changes to evidence and human remains, giving false or misleading information. Evidence exposed to open flames or gases may take up or release chemicals that interfere with many forensic techniques. However, despite the most unusual circumstances, forensic scientists can still pull pertinent information from such evidence, as long as the circumstances of collection are passed along with the exhibits.

With detailed notes and informed first responders, the next people in the crime scene analysis scheme—the crime scene investigators, the criminalists or scene-of-crime officers, and the forensic scientists—can begin to understand and interpret the evidence passing through their hands.

The Crime Scene Investigator

Fans of the popular television series *C.S.I.: Crime Scene Investigations* would likely be startled to discover that there is no such creature as a C.S.I. Level III—or Level I or II, for that matter—employed by the Las Vegas Metropolitan Police Department. There *are* criminalists (levels I and II), crime scene analysts (levels I and II), document examiners, evidence custodians, firearms specialists, firearms/toolmark examiners, forensic laboratory technicians, latent print examiners (I, II), and photo technicians. Of crime scene investigators, however, there isn't a hide, trace, or hair.

Job titles in the forensic fields can vary widely from jurisdiction to jurisdiction.

In Las Vegas, the job of a crime scene analyst is to "respond to crime scenes and perform a variety of investigative tasks to document the crime scene, including taking photographs, recovering evidence and processing latent fingerprints; and to perform a variety of tasks relative to assigned areas of responsibilities." The crime scene analyst is a civilian employed by the police department and (once again contrary to the television series) not a swab-wielding cop. He or she has no more authority in law enforcement than any man or woman on the street.

"This is the real scoop in Las Vegas. Years ago our CSAs, as they're called, were all commissioned personnel. They were essentially police officers who were assigned to the Crime Scene Investigative section. Since that time, in order to save money, as well as for other reasons, we started to civilianize the position until all of our CSAs were what we call noncommissioned personnel," explains Lieutenant Rick Alba, Director of the SCI Section of the LVMPD. "At first, our civilian CSAs didn't carry weapons. As time went on, we saw a need for them to carry them, but the organization didn't make it mandatory. What we ended up doing was allowing them to carry a department-approved weapon as long as they could qualify. Because it isn't mandatory, they are required to pay for the initial shooting course, pay for a concealed weapons permit, and they must stay certified by attending quarterly department qualification and training. They are allowed to carry guns just like any other civilian. They are also allowed to carry them concealed if they choose. But they must stay qualified. Additionally, they are not commissioned police officers; therefore, they do not have the same arrest powers that police officers do. If they had to make an arrest, they would have to do it just as any other civilian, by citizen's arrest."

At the time of this writing, the Albany Police Department's Forensic Investigation Unit in New York, while also designated a "support unit," consists of eight detectives—five on the day shift and three others on two overnight shifts. Its mandate reflects the multifaceted nature of these *police* investigators: "At a crime scene, every law enforcement officer shares the responsibility of preserving and collecting as much pertinent physical evidence as possible." Still, there's a separation between law enforcement officers and what the Albany PD terms "forensic scientists." For example, concerning DNA, the unit notes, "This type of evidence is left to the forensic scientist. What we at the Forensic Unit need to know is how to recognize, collect, package, and store the physical evidence for DNA analysis."

California's LAPD and LASD (Los Angeles Sheriff's Department) share many resources with their Scientific Investigation Division (SID), which operates out of the Support Services Bureau. Within the SID are eight sections: Criminalists, Photographic, Latent Print, Electronics, Questioned Document, Polygraph, Valley, and Firearms and Explosive Duties. Personnel can be civilians or sworn law enforcement officers, though, at the time of this writing, proposals were underway to consolidate these specialists in a Regional Forensic Science Center. About 1,800 people make up the various reporting teams—a very different arrangement than the five-person shift in Albany.

Obviously, concentrations of particular crimes require more, or specialized, people to investigate them. If you want to clear a lot of burglaries, fingerprint and trace evidence personnel are essential. Homicides require those skills as well, but they also often require DNA studies, blood splatter analysts, firearms experts, or toolmark studies.

Clearly, "crime scene investigator" is a job title that encompasses a wide range of skills and responsibilities.

TWO

WORKING THE SCENE: THE EVIDENCE

Friction Ridges

Forensic investigations aren't cheap.

Applying every known forensic test to every basic break and entry would quickly run criminal investigation budgets into something approaching the national debt, but even the smallest police force can afford soft brushes and a bit of black dusting powder. Unlike trace evidence, a fingerprint is unique. It can't be stolen, borrowed, or forgotten. If it's there, so was the person to whom it belongs. And prints are easy to understand, even by jurists who've never been inside a courtroom or a lab. Cheap, immutable, definitive—they're the sort of evidence every detective wants to have in hand when a case goes to trial. No wonder fingerprints are major exhibits in so many trials, and that their collection, filing, and retrieval are integral aspects of crime scene processing.

What is surprising is how often criminals obligingly leave a print behind.

Fingerprints adorn 2,000-year-old contracts in China, so detectives aren't exactly taking unfair advantage of their quarry. True, knowledge of a fingerprint's unique nature doesn't appear to have surfaced in Europe until the mid-1700s, and it took another hundred years or so before three important bits of

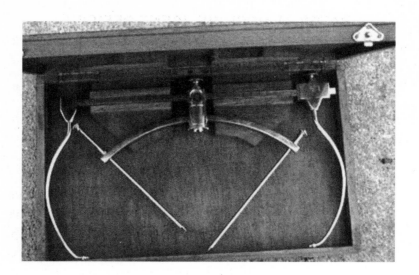

information all came together: the prints you have at birth are yours for life, no two individuals have the exact same pattern of prints, and, when properly catalogued, prints can be retrieved, regardless of how large a database you have. That last bit is extremely important. Earlier individual identification systems like Bertillion's prisoner measurements process (which took a dozen or so body measurements in order to catalog individuals) were abandoned because it was too difficult to search through thousands of files by hand to find a person with one arm a centimeter longer than the other.

In 1900, Sir Edward Richard Henry created a classification system that is still being used in the United States today, and (if one system isn't enough) Argentina's Juan Vucetich created an equal—if not better—method of cataloging that is also in widespread use today. With the basic process of storing and retrieving the relevant data established, it's been the areas of raw data collection and information sharing that have been the leading edge of the field.

PATRICIA MCGUIRE, IDENT OFFICER

"In one sense, every case is odd. Predicting what individuals will do before, during, or after a crime is possible to some degree, but there are always curve balls waiting."

For Pat McGuire, it was the scene of a multiple homicide that presented her with the most conflicting evidence she'd ever seen. "Collecting print evidence has become such a routine thing that, most of the time, you don't really have to be in ultra-observant mode. I mean, at most scenes you either will or won't find prints, depending on whether the crime was premeditated, whether the perpetrator wore gloves or not, whether—for whatever reason—they took off their gloves. If you have prints, you collect them; if you don't, you're disappointed but just move on to the next stage of documenting the scene. You don't expect to find prints you simply *can't* identify at all!"

Pat's crime scene was covered in prints—literally hundreds of them—but she knew she'd turned up something truly odd when, with the first swipe of her brush, she found herself looking at the four-inch-long print of a finger *tip*!

"It was clear as a bell—and clearly impossible."

Even though it was three in the morning, and she'd worked on many bizarre major-crime scenes previously, Pat wasted no time getting her supervisor out of bed. "I continued collecting the prints but felt completely out of my depth. Having stared at literally thousands, probably tens of thousands of prints by that time, I'd never seen prints this size. And the inconsistencies kept turning up."

By the time her supervisor arrived, still in her pajamas, Pat had managed to put together a coherent assessment of what was wrong with the prints.

"First of all, there was no side detail to any print. In the normal course of handling objects—and this was a messy scene, lots of things thrown around, overturned, broken—you'll see lots of prints showing the side aspects of a finger impression. These didn't. The central area was often clear, but the edges were always blurred. In the sixty or so prints I already had lifted, not one showed side detail."

Unusual, maybe, but possible.

"Then there was the layout of the prints. Prints turn up in predictable ways. If you've got a finger on one side of an item, you'll usually find the thumb on the opposite side. As far as I could tell from the prints already collected, the perpetrator grasped everything on one side only. I hadn't found one set of opposing prints, not one combined thumb and finger set."

Definitely weird.

"Looking at the damage around the room, not to mention to the victims, you were immediately struck by the sheer havoc of this scene. Three adults were dead: a twenty-one-year-old woman and both her parents. The younger woman was stabbed fourteen times, likely indicative of rage and a fairly personal rage at that. The parents were each stabbed numerous times and, even at that early stage in the investigation, it appeared that the mother might well have been tortured prior to death." Patricia still swallows hard as she looks back over the case materials. "The damage to property always seems insignificant compared to the human damage, but, just look at this picture."

It's the kitchen. Although all the victims had died on the upper story, either in or near their bedrooms, the kitchen might have been the scene of a minor explosion. Cupboards and drawers stand open with their contents smashed or thrown to the floor. Even the refrigerator has been yanked out of position, its contents spilled onto the floor.

"It's not unusual for enraged killers to attack things other than their victims. I've seen slashed beds and clothes, items that are intimately associated with the victim. Cars seem to come in for their fair share of abuse, as well. Computers, too, of late. A lot of jealous husbands think their partners are into chat sex these days. Guess that's a sign of our times, but in eighteen years I'd never seen anyone attack their victim's Corn Flakes or bargain-basement china!"

The scene was disturbing even to experienced officers like Pat, but to one neophyte investigator it was overwhelming. "The bodies were enough to make grown men think twice, and when one of the juniors sort of jokingly suggested the place looked like a gorilla had gone berserk inside, I think—just for a second—that everyone really stopped and wondered. It was just gallows humor, but. . . ."

Might explain the four-inch fingerprints?

"Yeah, I think that went through everyone's mind. But, while I'd never seen prints like these before, I'd also never heard of a real case where the murder weapon was a trained gorilla, so, well, I guess I just went in a different direction. Bizarre things happen on TV or in the movies, sometimes even in the real world, but this was a quiet residential area. Even at eleven at night—which is when we estimated the murders had occurred—there's no way you're going to walk a gorilla into and out of a home on a well-lit street without *someone* noticing it!"

However, the juxtaposition of gorilla images in her head with the actual scene in front of her suggested something almost that bizarre.

"My [supervisor] and I had worked a case a couple years previously where some elaborate staging of bodies had occurred at the dump sites, and, with that in mind, she and I walked through the scene again. The more we looked, the more it seemed too unreal, too violent. Even in the type of crime scene where violence would be expected, this was over the top. Put that with the outsized prints and you've got to start wondering if you're not looking at one huge stage."

Incredible as it sounds, Pat and her supervisor, Carla Tucker, began wondering if their younger colleague hadn't articulated precisely the impression the perpetrator had *intended* to leave at the scene—that something not quite human was responsible.

"People try to hide murders inside arsons, or robberies. Why not an animal attack?"

Later, during a slow walk-through with one of the detectives, Pat noticed that several items began to appear in a slightly different light. And ironically, it *was* a light, an overhead light, that began to firm up the conviction that someone was playing an elaborate and terrifying game. Everyone entering the scene had noticed the living-room ceiling fan dangling from its fixture, but no one had been able to place it within the evolving sense of what had happened at the scene. "We see that often in suicides, or attempts to stage murders as suicides, that a lighting fixture or showerhead will be torn loose," agrees Det. Ray Drover. "But, in the context of a multiple murder, especially one that had occurred upstairs, it just didn't fit."

The trace collector nixed the whole animal invasion idea immediately.

"A gorilla is larger than several big dogs, and they probably shed just as much. Yet, despite all the movement required to thoroughly trash the inside of that house, there wasn't any evidence of animal hair."

Hauling out the prints taken so far, McGuire, Tucker, and Drover began noticing more "consistent inconsistencies."

"Prints vary a lot. Not in actual features, of course, but in how they're laid down. If you're pushing something, prints get distorted in one direction. If you're leaning on something, they're almost always going to reveal more side detail than if you're picking up a light object. If you throw something, there's often a particular kind of blurring to the print, a wave pattern for want of a better description. Those prints are usually useless for identification, but, you can tell if something was dropped, tossed, or thrown by that pattern of smudging. Here, every print looked more or less the same, very little variation."

Carla Tucker holds up a copy of one print with the scale showing its unusual size. "In a way, picturing someone going around this house, with three dead people upstairs, deliberately making it appear that some creature had swung from the lamps and destroyed the place, was creepier than the simple fact of the murders. And, the longer we were there, the more we realized that's exactly what had happened."

The clincher?

McGuire taps the photo. "If you know there really was no gorilla or other animal, the next question is 'Where did the prints come from?' "

Model gorilla hands?

She nods. "But . . . ?"

Tucker speaks up. "Gorilla costumes, even good ones modeled from latex hand castings, don't leave *prints*, impressions perhaps, but not *prints*. Basic printing 101. You need oils of some type to do that."

"The premeditation kicks you in the stomach," adds Pat. "Today, a lot of people know something about criminalistics. It's in films and books, on TV almost nightly. It's still uncommon, but not unheard of, for criminals to deliberately try to foul the forensic evidence. But this. . . . This perpetrator had to have walked around that home, having already killed the people upstairs, having torn the place to pieces, thinking, 'I think I'll pull down that lamp. Gorillas swing, right?' And remembering to rub those rubber hands across his own face every now and then to make sure he could leave prints. Say what you like, that's cold."

As often happens, however, the more elaborate the staging, the more opportunities the perpetrators give investigators to track them. Pat McGuire's oddest case ended abruptly when a local costume company provided the credit card number of a man who'd rented a gorilla costume the day of the murders.

"Confronted with our reenactment, the guy just crumbled. The interview lasted about fifteen minutes. He simply couldn't believe we wouldn't accept what the evidence should, in his mind at least, have indicated. Crime scene reconstruction isn't about one piece of evidence—it's the whole scene that tells us what happened."

Tucker grins. "He should at least have gone to the zoo. He'd have known that gorillas won't swing in confined spaces. And they certainly wouldn't swing off ceiling fans. And anyone who has ever smelled a gorilla cage would immediately realize there's no way that odor would clear out before someone got to the scene!"

Prints themselves, of course, haven't changed. But the science of recovering them moves forward every day.

Criminalist Kari Day-Wells recalls, with justified satisfaction, "In 1998, I managed to raise a palm print on the bottom of a pool. Ten years ago, I wouldn't have even tried. And twenty years ago, even if I had gotten that palm, I'd have had nothing to compare it to because no one was collecting anything but finger tip prints, anyway."

Today, ridgeology involves the collection and analysis of prints from any section of volar skin: the skin of the palms, fingers, toes, and foot soles. In some cases, lip prints or ear prints are also collected by print examiners because the process of preserving these prints requires similar skills. But, as we'll see later in this section, most jurisdictions regard non-hand or non-foot prints as still being at the research stage.

Of volar images routinely collected and accepted by today's courts, there are three basic categories of friction ridge prints: patent, latent, and impressed.

Patent prints sport details easily visible to the naked eye. While this may be true of any print—surely we can all attest to the visibility of prints left on black appliance fronts—patent generally refers to prints made observable by extra substances, which coat the skin and are transferred to some object. At crime scenes, blood may create visible prints. So can oil, ink, catsup, and dozens of other substances. Because of the foreign coatings, however, many perfectly patent prints can't be collected in the usual ways. On the other hand, they may be so visible that photos alone can be used as the primary collection tool. So, even obvious prints must be evaluated individually and may need to be moved to the lab before a lift (a collection of the material making up a print) can be made.

COLUMBO'S HANDKERCHIEF

While most modern television programs at least make a nod to getting the science and procedures right, a few notables seem to delight in making their audiences scream, "What **is** he *doing*?"

My personal favorite television detective is the rumpled Columbo, always quick to whip out his equally rumpled handkerchief in an effort to keep his prints off candlesticks, ashtrays, wine bottles, or, in

"Grand Deception," a two-foot-long black metal flashlight with a highly polished surface! Not only did he grab the light with that grungy hankie—wiping it as clean as any crook would have done—he chose to hold it by the handle, the very spot where his poor lab technician would first look for prints.

Latent prints—those not generally visible to the naked eye—result from the transfer of normal oils and salts from the skin to some surface. Like the prints on household appliances, many latent prints pop into view in the right light. Others need augmentation by powder or chemical treatment.

WANTED–LATENT PRINT EXAMINER

The Montgomery County Department of Police is seeking applicants for the position of Latent Print Examiner.

Qualifications include completion of high school; formal training in both the classification, searching, and filing of inked fingerprints and comparison; three years of experience preparing, lifting, comparing, identifying, and preserving latent prints and related evidence; and three years of experience taking, comparing, and identifying fingerprints.

Responsibilities include comparing and identifying whole and fragmentary latent prints lifted from or developed on various and possibly unstable surfaces; testifying in court as an expert witness; using the Henry System of fingerprint classification; and classifying, searching, and identifying fingerprints accurately.

Salary: $33,699–$55,619/annual.

The third type, impressed prints, aren't transfers of any kind but actual physical moldings of a set of friction ridges. Caulk around windows, wax, gum—almost any malleable surfaces—have the potential to contain print impressions. On top of functioning as a natural ink, blood can also become a perfect substance for picking up impression prints as it solidifies. Kari

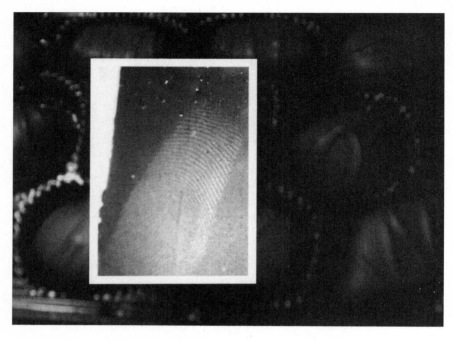

Day-Wells, working an apartment, once discovered the session's best print on a Christmas chocolate!

"This had been a nightmare crime scene—no nonresident prints anywhere. Then, there it was—this beautiful complete thumb on a piece of chocolate in a Christmas candy box! It's funny how we think sometimes. The suspect touched nothing else, but was caught because he didn't like nuts. God bless chocoholics."

In general, photography captures these impressed prints, but reverse molding or casting may be useful as well. In some labs, like Day-Wells's, impression prints are called "plastic prints," but either term refers to a 3-D print pressed into a surface.

MANNI GOMEZ

Manni Gomez has collected impression evidence in some pretty odd locations. And after twenty-two years, he has come to the conclusion that a criminalist must learn to apply lessons learned on past jobs to the situation at hand, regardless of how dissimilar the scenes appear to be.

"I worked in Canada for a bit back in the early 90s. That's where I discovered that obvious tire prints—apparently fresh and left in the snow just moments ago—can actually be artifacts of *much* earlier events. Luckily for me, the people I was working with—all RCMP trained—were more than happy to explain what they called "The Yeti Effect" to a sand-hopper like me.

"Apparently, an anthropologist was investigating the Abominable Snowman legend with several others in either Nepal or Tibet, I forget which, when a local appeared on their doorstep to let them know that, if they wanted to see some Yeti tracks, they should hightail it up the side of yonder mountain and have a look before it snowed again. They did and, sure enough, were directed to what appeared to be huge footprints separated on a slightly-larger-than-man-size walking stride. Three of the four set up their cameras and started documenting the impressions, but one of them, who'd grown up around snow and figured that snow was snow wherever you found it, started walking along the track. It was a bright, sunny, but cold day, just like the ones he'd known growing up. Eventually, he found a shaded spot and, sure enough, discovered that those huge single footprints had started life as four closely oriented individual prints. Snow doesn't *have* to melt into puddles before it evaporates. Under the right conditions, with everything still cold, but good steady sunshine overhead, snow can go from the solid form to the gaseous almost directly. The process itself is called 'sublimation.'

"The implication for those of us collecting tire impressions from snow is that, while an imprint might appear to be newly laid on a fresh surface, it's entirely possible, under the right circumstances, that the 'surface' print is actually a remnant revealed when the real top layer of snow has disappeared on a crisp but sunny day. I once spent hours retrieving impressions that had probably been there since the fall. When one of the guys who'd trained up there saw what I was doing, he walked me along to a shaded area of the tire marks. It looked like they just disappeared, like something had reached down and plucked the car right out of there. Absolutely pristine snow spread out from there, and there wasn't a mark to be seen.

"There weren't really any aliens, of course. When he took a bristle brush and started sweeping the spot where the tracks appeared to end, you could just make out the imprint continuing *under* that layer of snow."

Handing over a narrow tire cast that still has grains of sand clinging to it, he adds, "Pity I didn't remember that idea when I came back to the high desert here in the Southwest!"

Called to the scene of a buried body, Manni was asked to collect nearby tire impressions. The assumption was that whoever had buried the body had driven to the scene—a perfectly reasonable idea, as the body was nearly forty miles from the nearest community.

"There were two different sets of impressions within normal walking distance of the burial site. One was clearly a van or a truck by the initial measurements of the wheel base. The other tracks had completely different wheel base measurements, ones I'd not encountered before, and, to my eye, the individual impressions looked more like mountain bike prints, but without the tread. Really weird!"

Aware that whoever had reported the body had probably driven here, Manni asked if anyone knew the make of the vehicle that person had been driving. It turned out to be a 2000 Ford Windstar. So, with the larger set of tread accounted for by the couple who'd stumbled over the body while looking for a camping site, he was left only with the one remaining set. However unusual the tread, this set was clearly the one most likely to be associated with the body, especially in a remote, unfrequented area.

The marks were clear enough. There just wasn't any detail in them.

"The measurements I took that day were of a 'tire' about four inches across, with odd repeats that showed some impression in the surface, but absolutely no tread marks. Even tires you'd describe as bald, have some tread marks. And these tracks, despite being an unusual size, were nice, clear impressions."

Looking at the cast again, it's easy to see what he means. The marks have sharp edges that are not at all blurred.

"I took the casts and brought them back to the lab but, frankly, had no idea what I was going to do with them other than file them and see if a vehicle would turn up later in the investigation."

As it turned out, there would be no investigation. In the process of excavating the body, it became clear that this was not a recent burial. It might even have qualified as an archeological dig!

Carrie Frumm, assistant medical examiner, recalls the case. "Bodies dumped in the desert can, when circumstances are right, be preserved by the dry conditions, the high heat, and the relative lack of predators. They aren't exactly mummified, but they don't always decay normally, either. The woman recovered on that occasion likely died over 150 years ago."

Says Manni, "The track I'd cast, the one you're holding in your hand right now, is a wagon track. There *was* no rubber to leave tread impressions!"

And how did a 150-year-old track come to survive in such pristine condition?

"Well, if sand was snow, I'd have walked that track just to make sure it didn't disappear somewhere along the way. But, as this wasn't snow, I didn't walk it looking for anything inconsistent. Thought I'd left all that sublimation stuff back in Canada."

But sand doesn't sublimate, does it?

"No, of course not. But, like snow, it gets laid down in layers, and each layer has a surface that is acted on by natural forces and elements. As near as we've been able to reconstruct it from going back and following the track, some 150 years ago the desert had one of its rare big rainfalls. Either during or shortly after that rainfall, this woman died. I don't think they ever found out how, but, 150 years later, there wasn't anything that could have been done about it, anyway. In any case, she was buried, like hundreds of others, when the traveling party stopped. She wasn't buried that deeply, likely because the ground *was* wet, or because there wasn't much time, or nobody had a decent shovel. Or maybe they just didn't care. The next few days must have been so incredibly hot that they baked the ground with the wagon impressions hard and warmed the ground enough to keep that woman from decaying in the usual way.

"Over time, wind, maybe even more rain, deposited new layers on top. Sometime in the early spring of 2001, wind or rain washed away that top layer again, revealing the hard-baked impressions below and part of the body—one foot.

"When we went back up there and actually followed those marks to lower ground, they disappeared, just like the marks in the snow had. Taking out my own bristle brush, I swept the edge and, sure enough, there they were under a layer of sand. The round marks were the ends of dowels or metal brads used to hold the wagon wheels together. Two completely different impression materials, two completely different environments, but the same underlying theory."

Despite the less controlled circumstances, printing done at the scene is often preferable to attempting transport of items. Even expert packaging can't guarantee that smudging or cutting (the formation of tiny scratches across a print) won't occur. Some prints simply can't be processed on site, while others simply can't be moved. Consequently, investigators must carefully weigh risks, benefits, and the technology required before beginning collection. Kari Day-Wells once took an entire doorframe and door back to the lab to avoid environmental problems because extreme cold was preventing powders from working. The next week, she spent over an hour ensuring the safety of one print in the dusty corner of a window in hopes that a photographer could light and capture it.

"Prints in dust can't be powdered. It destroys them. So I overturned a glass on it and waited until the photog could break away from primary documentation in another room."

"Every scene is different," says Day-Wells. "One of my instructors tells of removing an entire wall because it was covered in old-fashioned, uncoated wallpaper and he wanted to fume it.

He turned up the only decent prints of the case. That was in the days before good portable fuming equipment, but I've had to do similar things as recently as last week. We have a lot more tools available to us now. We'd be fools not to evaluate prints in terms of both field collection and lab testing."

A selection of powders (including fluorescent varieties that react when illuminated with alternate light sources, or ALS) and magnetic types (used with a magnetic wand instead of a brush) can be used on nonporous surfaces in the field with good results.

Lab testing, however, allows time to evaluate a print's composition and the surface on which it appeared. This surface could be highly porous or absorbent like paper, or the shiny, nonporous finish of a firearm. Many chemicals pose significant health risks, making field work with them dangerous. In fact, federal and state regulations on hazardous substances may well make field use of some chemicals impossible. Transferring evidence to the lab also allows print examiners to approach printing in a multidisciplinary way. Some methods of raising prints on a firearm, for example, make later ballistics testing impossible. For example, it's virtually impossible to get a result from iodine fumes if silver nitrate has already been used. And iodine results, frequently short-lived, may require a photographer devoted to capturing results as they appear—easier to accomplish in the lab than the field. Being able to work out a total plan for any piece of evidence has obvious advantages.

Some printing techniques usually undertaken in the lab do make their way into the field when circumstances demand. Fuming can take hours to yield results on large items (such as an entire car) and only works in an enclosed environment, so it's often a lab technique. Cyanoacrylate fumes, however, can stabilize many delicate prints, so portable fuming chambers are used by many labs. It's not unusual for prints to receive a preliminary processing in the field, and to be treated to dyeing or other techniques back in the lab.

FINGERPRINT EXAMINER'S FIELD KIT

Every examiner's kit varies. In fact, many crime scene investigators have several different kits prepared—one very basic and one with the more exotic tools so that those materials can be kept handy but don't have to be lugged out for all occasions.

- Powders—Regular powder generally comes in black, gray, or white. Powders of exotic colors are produced commercially, but generally they are black or gray, or sometimes white—whichever will contrast best with the print surface.
- Fluorescent powders—A difficult or multicolored surface may require special handling, including fluorescent powders that show up in combination with an ALS or UV light source. Depending on the light source, special goggles might

be necessary, as well. Lifting and/or photographing these prints often requires special procedures.

- Magnetic powders—Used in conjunction with magnetic brushes or wands.
- Cyano products—Chemical print developers are usually reserved for the lab, but cyano products (such as Super Glue) and a portable rig for containment may be included.
- Lifting tape—This is not household tape, though household tape has been used in emergencies. It is a wider, low-tack adhesive strip not affected by oils and other chemicals.
- Rubber lifters—These conform to uneven surfaces or curves better than lifting tape. Unfortunately, these lifters, which aren't transparent like tape, show the print in reverse so you may have to print the lifter itself, either photographically or mechanically, to flip it back to the correct orientation.
- Cards on which lifting tape can be affixed—As the most commonly used powder is black, most cards are white for contrast, but black cards are usually tucked in a larger kit, as well. Commercial applications in which the tape and card are one unit are also available.
- Brushes—Some prints require the most delicate touch, so all examiners have a variety of brushes—some natural bristle, others fiberglass, feather, or camel hair.
- Printing ink, roller, preprinted ten-finger cards—This is the usual gear for taking prints from people. Living people can and usually are printed by bringing the person to the examiner. Deceased individuals, however, are usually printed during the autopsy exam and the examiner may need to go to the body.
- Flashlight and batteries
- Camera, flash, optional tripod, extra film and batteries—Special print cameras exist and some jurisdictions use them; others use standard 35mm equipment or Polaroid

systems. Many are switching to digital systems. A video camera is also becoming standard gear, even for examiners, as it helps to fix print locations within the larger scene.

- Disposable petri dishes—These are not standard items but some examiners have found them useful in protecting prints at the scene. Inverting one over a suspicious spot warns other investigators to use extra care.
- Personal protective equipment (masks, booties, gloves, etc. as required)
- Casting or molding supplies, mixing containers
- Distilled water
- Tweezers, chopsticks, or other tools for manipulation
- Ruler, compass, sketching materials
- Pencils, markers, pens

Once there is a visible print to work with, the next step is to recover it and to document its location at the scene. Photography is a powerful tool for both capturing an image and fixing its original location in the record. Many jurisdictions make it policy to photograph every print. But the more common way to recover a print is to create a lift—actually removing the material adhering to the print and affixing it to a card for permanent storage. The process, though straightforward, requires a delicate touch and attention to detail.

- Being careful not to touch the bristles of the brush (powder will adhere to oils on the brush as readily as to oils in the prints, creating a clumpy mess), fluff the brush by gently shaking or twirling it between your palms.
- Use a light touch on the first stroke across the latent. Pause to see which way the print is running before gently continuing to build up powder by stroking in the direction the print is going. Excess pressure or strokes across the line of the print can cut it, creating streaks.
- Blow off extraneous powder. (If this scene requires personnel

to be masked inside the tape, a can of compressed air can be helpful.)

- If appropriate, photograph the print.
- Apply print tape to the entire print, starting at the top of the print and moving steadily. You need complete contact, no wrinkles or "fish eyes" (bubbles). With tape in place, firmly press straight down on the entire surface. Don't rub it about, just press down. (Like the art of dropping a cover on a microscope slide without trapping bubbles beneath, taping skills improve with practice.
- Just as placement requires a steady hand, removing tape from the surface without wrinkles or creases requires a smooth lift without pauses.
- Carefully lay the tape on the card and ensure there are no loose corners.
- Document the collection. Immediately initial and neatly jot notes directly on the card and in the examiner's notebook, making sure to have each lift numbered and mapped to the larger scene maps and diagrams.
- Repeat if it appears a second lift might be better than the first.

If a good lift is available, the fingerprint technician begins the examination and comparison phase of the investigation. If a second print—from another scene or a suspect—is already available, a direct visual inspection of the two prints begins. If not—if the print will have to go into the database in search of a match elsewhere—then the examiner must also prepare a description of the print based on the minutiae of the print.

> In an early Paula Palmer novel, the detective returns from the fingerprint lab delighted to have found "a ninety-five percent match!" In reality, there's no such beast. A print is either a 100 percent match or a 100 percent mismatch. There's no in between. Because prints are unique to each individual, even the slightest difference between any two prints means there simply is no match.

EIGHT PATTERNS IN PRINTS

arch: This looks like a wave across the print. The wave is smooth, rounded at the top, without points.

tented arch: These arches, sort of exaggerated versions of the plain arch, push up sharply in the center of the wave, forming distinct triangles at their centers.

ulnar loop: A loop in fingerprinting is like a loop anywhere else, the ridges appear to flow in from one side of the print, turn back on themselves, and leave on the same side from which they entered. Ulnar loops are loops that flow in the direction of the little fingers.

radial loop: Radial loops are loops that flow in the direction of the thumbs.

whorl: The plain whorl looks much like an eddy in water, otherwise known as a "swirl." Technically, an examiner would

	say that at least one ridge in the pattern makes a circle-like shape: circle, oval, or spiral.
double-loop whorl:	This whorl is formed when two loops appear to run into one another and swirl around in a circle. Two deltas (small, triangular formations or small dots that occur where the linear flow of a ridge is disturbed) mark the bases of the two loops.
central-pocket loop whorl:	This is a feature within a feature—the loop rests in the middle of a whorl, interrupting the outer edge of the whorl as a delta.
accidental whorl:	The accidental whorl is defined by the parts of the whorl that are or are not inside lines drawn between deltas. In practice, most accidental whorls often appear as small whorls inside what would otherwise be a tented arch—sort of a "pointy whorl."

Analysis of a print begins by determining which pattern describes it. As all prints fall within one of those patterns, each of the eight fingers and two thumbs generates a description based first on whether or not the print is whorled, the type of whorl, the number of whorls, etc. Each yes/no response generates a number in the examiner's code and each numerical response, such as "three whorls," can also be translated to a set numerical description. The complete code exactly describes the print or prints in a way that machines—and people—can understand. As print databases grew well into the millions, making a match without a machine that speaks the same language as the examiner became almost impossible. The current FBI system, based on Sir Edward Richard Henry's work, can take those numbers and present examiners with a manageable list of possibilities.

In order to keep the databases held at all levels of government up to date, every print, whether submitted to the databases for matching or used in direct comparison in-house, should be classified and added to the appropriate section of the Automated Fingerprint Identification System database, or AFIS.

While the search engine has been automated and computers are doing the grunt work once undertaken by examiners, classification (information going into the system) and comparison (evaluation of information coming from the system) remain essentially human functions. It's humans who sit in the lab and stare at each print, identifying landmarks in the mass of swirls and ultimately declaring a match or mismatch. For that verification, examiners need to get a little more intimate with a print.

A restaurant named El Pico used to serve a dish called Ultimate Spaghetti. It came to your plate with the end of one noodle already wrapped around a fork and, if you were a very good slurper, you could suck up the entire dish without touching the fork more than once. It was actually all one long noodle. Kids loved it, even if their parents weren't so keen.

At a glance, it might seem like the prints on your fingertips have a lot in common with Ultimate Spaghetti—one big snarl that's really hard to see as anything except a continuous line. If you really look at them—try it, get yourself a clean glass, press your fingers to it, and hold it up to the light—you'll discover there's no line clearly twined around a starting point. Instead, there are dozens of tiny breaks, forks, and swirls hidden in those prints. The ridges, or raised areas, and the furrows, the spaces between ridges, act like the traction treads on your tires, allowing you to grasp objects securely. If the skin on your hands was completely smooth, lifting a rock wouldn't be much more difficult, but turning the page of a book would get cumbersome pretty quickly. And to a print examiner, those aren't just squiggly little lines. Each possible ridge type falls into well-designated classes of minutiae.

- **bifurcation:** occurs when a ridge splits or forks into two separate ridges

- **ridge end:** the abrupt termination of any ridge
- **lake, or enclosure:** occurs when a single ridge line bifurcates, then quickly reconnects and continues as a single ridge
- **island:** short ridges that simply start and stop and don't connect to any other ridges
- **dot:** a very short island, of basically equal length and width
- **bridge:** a short ridge that runs between two other parallel ridges, sometimes called a "railway tie"
- **spur:** a special bifurcation where one resulting ridge is considerably shorter, ending near the original split or fork

Each set of ten prints contains differing numbers of each of these features. Personally, I don't have any dots in my prints, but I have dozens of lakes on just one of my thumbs. Some features occur more often within populations, as well, but the percentages on any particular feature's chance of occurring are meaningless. Examiners don't think in any percentages except 100 and 0. For example, my having forty-two lakes on my left thumb doesn't make me a 20 percent match with Mother Teresa, who also had the same number of lakes on her left thumb.

As print recovery methods dramatically improve, more comparative techniques—previously only theoretical—have begun to receive more attention. The field of ridgeology is expanding.

Edgeoscopy is the study of the features found along the edge of *individual* ridges. Going from ridgeology to edgeoscopy is like switching from Sherlock Holmes's magnifying lens to the electron scanning microscope. Because field prints generally fall below the level of clarity required to study such tiny features, however, work continues with whole prints collected directly from subjects onto print cards or, as is happening more and more often, scanned at high resolution directly into a computer system.

Fingerprint Keys

Scanning technology itself has pushed print identification forward, as well. Digitization allows sophisticated software to produce a virtual map of a print, identifying the minutiae of ridges, terminal ends, bifurcations, and other features. The size of the print, which changes as we grow from infants to adults, is immaterial. Distances between points are relative to the entire print, not measured in a particular scale. For example, a dot might be found twice as far from a particular ridge as an island. That relationship of two-to-one doesn't change, regardless of how big the print becomes over time.

Print scanners, once used only on the entry systems of bank vaults, now come as a security feature on laptop computers being lugged all over U.S. college campuses. Livescan, a print scan system and database proven capable of collecting high-quality prints from individuals, doesn't yet include anything except fingertips. As nearly a third of the hand prints taken from scenes are palm prints, this gaping hole in the new system will have to be addressed if it is to be as useful as the older systems, but it certainly shows great promise. Similar biometric technology may soon be built into car ignitions and home lock sets.

So much for remembering passwords!

Poroscopy, like edgeoscopy, is also concerned with tiny print features that may be visible in a scan or quality print card. But here the focus is on the pores that secrete oils and salts from the skin. The idea that patterns among the pores would be as unique as the ridges where they were found was first proposed in 1912 by Dr. Edmond Locard, often called the Father of Forensics. Like edgeoscopy, however, poroscopy couldn't be put to practical use until print collecting techniques caught up with the theory.

Both poroscopy and edgeoscopy undoubtedly hold great promise for forensic and identification applications in the future. But for now, the print characteristics in use for the past 100 years

continue to prove their worth in modern, computerized investigative work.

Prints captured as lifts with non-fluorescing powders, or as image scans, aren't the only sources for comparison. The cutting edge of print capture moves ahead almost daily, and one of the field's best developed tools is chemicals.

Applied either by direct means like immersion or less direct means like sprays or gases, chemicals react with components in prints to create a visible combination of the two. These prints aren't lifted like powdered prints, but photographed.

The most common chemical treatment in printing non-porous objects is fuming with Super Glue, a cyanoacrylate. The fumes—obtained by simply heating the glue to its boiling point (just over 120°F) inside a confined space—float over the object being printed. Chemicals in the fumes adhere to the amino acids—proteins and fatty acids—in the print, building up a white substance that can be photographed directly if the object is white, or further treated with black or colored powders or dyes and then collected. The stickiness of the glue makes it a good powder subject and a number of fluorescent dyes specifically designed for cyano-prints are also available.

When Bosch and Billets stepped back into the shed, Donovan was at one of the worktables working with the leather jacket. He had hung it on a wire inside an empty one-hundred-gallon aquarium and then dropped in a Hard Evidence packet. The packet, when broken open, emitted cyanoacrylate fumes which would attach to the amino acids and oils of fingerprints and crystallize, thereby raising the ridges and whorls and making them visible and photo-ready.

MICHAEL CONNELLY, *Trunk Music*

Humidity, concentration of fumes, size of sample, size of fuming tent or chamber, and any number of other factors can affect reaction time, but overprocessing can make one big, ugly blob of the print, so the process requires supervision.

Reaction time can be shortened by putting small fans inside the fuming chamber, which moves the fumes about, allowing more fumes to reach the print faster. Mixing sodium hydroxide with the glue produces lots of fumes even without boiling. Also, just as low humidity can slow the fuming process, increasing humidity in the fuming chamber can accelerate it, so it's common to add a container of water to the heater.

SPECIAL AGENTS FOR SPECIAL PURPOSES

Iodine Fumes

Cyanoacrylate isn't the only fuming substance useful in print examination. Iodine crystals that have been heated to release fumes will react with oils on porous surfaces like paper, producing a yellow image. This is a favored agent for espionage work as the print will fade again and papers can be returned without leaving evidence of having been printed. Of course, this does mean that the fuming must be closely watched and there should be a photographer ready to take images as the print develops!

Iodine Reagent Spray

A spray that can be used on both porous and nonporous surfaces, this is actually two solutions that are mixed at the time of use. The first is the iodine dissolved in cyclohexane; the second is a-naphthoflavone and methylene chloride. Once mixed, the spray is good for about a day. It should be applied sparingly for best results.

Gentian Violet

Once the standard for prints on sticky surfaces like tape, it has been largely displaced by a commercial product called Sticky-Side Powder, which is mixed with a film cleaner called Photo-Flo, or in-house preparations made of soap and black powder. A mix of Photo-Flo and ash gray powder works well on dark surfaces like the adhesive side of electrical tape. The newer mixtures all come with the same basic instructions: paint on, wait for about a minute, rinse with cold water, photograph. But gentian violet is a dipping solution; objects are submerged for a minute or so, then rinsed.

Liqui-Drox

For a dye useful on both sides of adhesive tapes, especially dark tapes, a combination of Ardrox P133D, Liqui-Drox, and distilled water can be used, as long as the non-sticky side has been cyano-fumed. This application is a little unusual. The mixture is spread on both sides with a brush and agitated until a foam appears. After letting it sit for ten seconds, rinse well and air dry. Prints will appear in long-wave UV light. These prints will fade again, so the process should only be undertaken if a photographer is ready. The prints can also be lost if exposed too long to UV light, so it's best to simply confirm their presence and contact the photographer.

Magnetic Powder

This powder isn't really magnetic itself, but it works with a magnetic brush or wand. Used early enough, it can make prints on papers visible. After a few hours, other methods must be employed. The process requires a delicate touch as the wand must not touch the surface, just approach closely enough to deposit or pick up the powder. With the magnet engaged, dip the wand in the powder container. Powder will adhere to the tip and can be "brushed" over the surface. Once the print is visible, the remaining powder on the wand can be deposited back in the bottle by holding the wand inside and removing the magnet. To remove excess powder on the print, replace the magnet and sweep gently above the surface.

Ninhydrin

One of the major components of prints is amino acids. Several substances react with them, but ninhydrin is particularly good, resulting in a bright pink and purple image, which is permanent. On the down side, it can take several days for some prints to appear if the process isn't helped along with heat and humidity, and the stuff is well known for its ability to cause blinding headaches in examiners. Application is easy enough: simply spray, or dip if running inks won't be a problem.

DFO

Like ninhydrin, DFO can be used on porous surfaces and reacts to amino acids. Under an ALS (alternate light source) or a laser, DFO fluoresces. The solution is made up of methanol, ethyl acetate, and glacial acetic acid and contains a small amount of DFO. It can be used as a soaking solution or sprayed but must be oven dried (or ironed) for twenty minutes.

Silver Nitrate

A favorite for examining a wide variety of papers, silver nitrate is typically a water-based solution. However, if the paper or cardboard has

a waxy surface that would leave the solution beading on the surface, it can also be alcohol-based, making it very versatile. Unlike iodine fuming, silver nitrate permanently marks most specimens—as well as the examiner if it touches skin! On waxy or glossy surfaces, prints should be photographed as they appear because they may not be stable for very long. Having a lot in common with photography chemicals, this solution is photosensitive and, after the items have been soaked or painted, they must be subjected to bright light to induce development of prints.

**Amido Black,
Methanol-Based Formula**

Prints in blood require special handling. This is a three-step process. Naphthol blue-black dye, glacial acetic acid, and methanol are carefully mixed in a magnetic stirrer or similar device for about half an hour, or until the dye is dissolved. Spray or dip the item, let it stand for thirty to sixty seconds, rinse in a mixture of glacial acetic acid and methanol without the dye, then rinse once again with distilled water. This solution should not be used on painted surfaces as those surfaces have a tendency to melt, ruining the prints.

**Amido Black
Water-Based Formula**

Also used on blood prints, the water-based chemistry is a more complex mixture than the methanol-based formula, but is safer on some surfaces. The process is only two steps long and doesn't require distilled water for the rinse. Once again, glacial acetic acid and naphthol blue-black are mixed, but instead of the methanol base, a combination of distilled water, Photo-Flo, formic acid, sodium carbonate, and 5-sulfosalicylic acid are added. Specimens are dipped or sprayed and allowed to sit for three to five minutes before being rinsed. This is a repeatable process.

DAB

DAB, another formulation for enhancing prints in blood, *cannot* be used after cyano-fuming, but works well otherwise, even if the process requires four steps. Solution 1 is a fixer of 5-sulfosalicylic acid in distilled water; solution 2 is a buffer of phosphate solution; solution 3 is the DAB itself; solution 4 is a developer, basically a mixture of solutions 2 and 3 and a little hydrogen peroxide. After soaking the item in the fixer for three to five minutes, submerge it in distilled water for about a minute, then transfer to a tray of developer until good contrast is achieved, in any case no more than five minutes. To stop development at any point, submerge the item once more in developer. If the item cannot be submerged, a series of tissue pads can be applied to the appropriate spot and solutions applied to the pads for the same time intervals. New pads must be applied between steps.

Leucocrystal Violet (LCV)

A spray dye for enhancing prints in blood, LCV, a mixture of hydrogen peroxide, 5-sulfosalicylic acid, sodium acetate, and LCV, should be finely misted, perhaps with an artist's airbrush. This mixture *cannot* be used after cyano-fuming and should be used only cautiously in sunlight as this agent can develop in bright light wherever it is sprayed—not just on the prints.

Coomassie Brilliant Blue (CBB)

A dipping or spray solution, CBB is used in a three-step process involving a developer of Coomassie brilliant blue, glacial acetic acid, methanol, and distilled water, a rinse of the same ingredients without the Coomassie brilliant blue, and a second rinse of distilled water. The developer is left on for about a minute, then rinsed. If the required contrast hasn't been achieved at that point, the process can be repeated. Use the second bath when the print is clear. Crowle's Double Stain—a Coomassie and Crocein Scarlet 7B combination—can be used in similar circumstances, with latent or blood prints, in the same way.

Sudan Black

An immersion solution made from Sudan black dye, ethanol, and distilled water, this is best used on surfaces made difficult by oils. It reacts to elements in perspiration and produces sharp blue or black images.

Ardrox

A fluorescing dye for use on colored subsurfaces, it must be viewed with a UV light source after spraying or dipping.

MBD

Another of the fluorescing dyes, this one serves well after cyano-fuming. It shows latent prints under either a laser or ALS. Dipping or spray bottles provide the best coverage for most items. MBD can be made by dissolving the stock in acetone, then further mixing with methanol, pet ether, and isopropanol. MRM 10 works as MBD on nonporous surfaces. Rhodamine 6G is a fluorescent dye that works well on colored surfaces. In both cases, a laser or ALS is applied after dipping or spraying.

Safranin O

Like other fluorescent dyes, Safranin O is useful in making cyano-prints visible under laser or ALS, but it really shines in the shorter lightwave ranges.

Vacuum Metal Deposition (VMD)

A technique completely unlike any noted so far, VMD makes latents on a variety of specimens visible by laying down a thin layer of gold and zinc on items inside a vacuum chamber. It should be noted, however, that it is not applicable to porous surfaces. Needless to say, this is not the most common method of raising prints!

So far, we've concentrated on the tips of the fingers. But, as already stated, the volar areas of the body include the entirety of the palm, the underside of the fingers from tip to base, the outside edges of the hands, and the soles and toes of the feet. Consequently, the old ten-print cards, which captured just the fingertips, quickly grew into the major-prints card. These include the rolled tips of the fingers (rolling involves inking the fingertip and then rolling the finger across the card to capture the whole ridge design, not just the bit in the center); whole hand prints that include the palm and edges of the hands (usually obtained by wrapping the card around a can or other container and rolling the inked hand over it so the palm presses firmly); and even, in some cases, the sole and toe prints, as well.

As this is all friction skin, the same as the skin at the fingertips, it is also unique to one individual and the expanded area of comparison makes it possible to apply printing methodology to

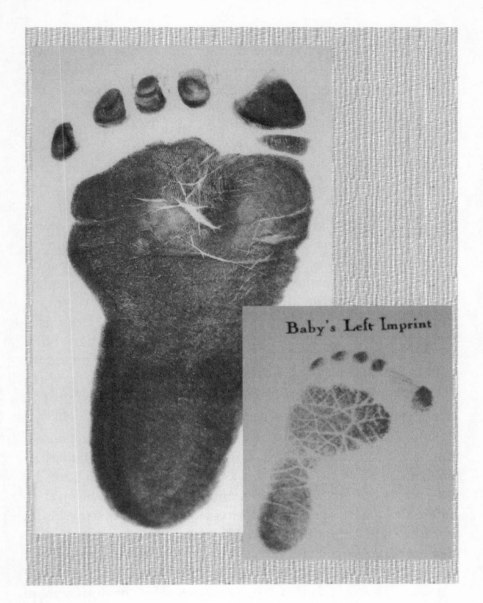

Baby's Left Imprint

the whole hand. The edges of the hand, for example, frequently come into contact with paper during handwriting, but the tips may never rest anywhere but on the pen. Being able to catalogue palm and sole prints greatly increases the likelihood of finding exculpatory or identification evidence. Though most hospital prints of infant feet fail to produce useable images, a few have played important roles in identifying missing children.

Not all skin is volar skin, however, and several other "skin impressions" and "skin prints" are actively being investigated as possible further additions to the print field. Ears in particular have been closely scrutinized for some years. It's been recognized among investigators that ear prints often turn up on doors, occasionally on windows, and even on walls. Apparently eavesdropping is popular with criminals, especially burglars who listen to confirm that a residence or other area is empty. Other items that leave impressions, like shoes, have been used in identification before, and presumably ears could fall into that category if there were actual three-dimensional impressions. Unfortunately, doors and walls are considerably harder than snow or sand and don't take impressions. And as prints, the status of ears is nebulous.

Ears don't have ridges, so it's not possible to apply the classifications used in ridgeology and edgeoscopy. Alfred Iannarelli published one of the earlier books on ear identification, but he was working from photographs and not prints, and had yet to subject his methodology to peer review or field testing. From time to time, it has been possible to use ears to *exclude* individuals from consideration. For example, the adult woman with attached earlobes can't be the lost child who had unattached earlobes. The *inclusion* of suspects, however, is obviously more difficult, and the *identification* of a unique individual appears out of range at this time.

One suspect in a convenience store burglary captured on video was identified in spite of the cloth tied over his nose and mouth by a large mole noticed just inside the outer rim of his

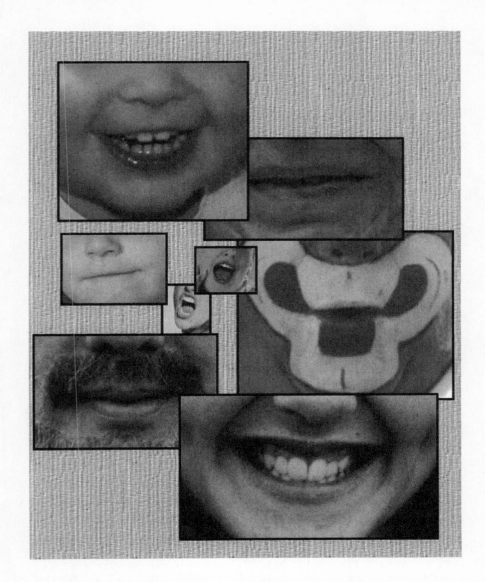

ear. He confessed, however, and the ear ID was never challenged in court, so we'll never know if it would have been accepted by a judge or jury as definitive evidence. There is currently no general acceptance of ear identification—let alone ear printing—in North American courts.

READING THE TRANSCRIPTS

While reading transcripts of a Frye hearing on the general acceptance of ear print validity within the scientific and forensic fields, I read this inadvertently funny line: "I think it's entirely probable that it's the defendant's ear that was found at the scene." Let's hope their suspect was a one-eared man.

Lip prints frequently elicit scientific curiosity as a possible means of identification. Left on cigarettes, glasses, even people, they're certainly numerous enough to warrant forensic notice, and lipstick can make many of them easy to find. Once again, however, there have been no peer-reviewed studies, there are no systems of classification, and no means of individuating one pair of lips from any other seems to have been developed. In fact, lip prints may present special problems in classification. Compared with the skin of the hands and feet or the even stiffer construction of ears, lips are incredibly pliable. The relative difference in reference points between a wide smile and a pucker is inches! Being able to account for such massive distortion would add but another step to the comparison process—a comparison process that doesn't even yet exist.

Still, while these non-volar prints aren't generally acceptable as identification evidence, the possibility continues to be investigated and the interest in such prints and impressions can yield other benefits. A 1984 California murder case might have passed unnoticed—and certainly unproven—if investigators hadn't found some time later the print of a woman's face inside a plastic laundry bag at her home. Without an awareness of

print transfer basics and the fact that perspiration can leave impressions on plastic, no one would have been looking at those surfaces in the first place.

With DNA being called the "new fingerprinting," there's been considerable public speculation about traditional fingerprinting becoming obsolete. But print examiners aren't worried about shrinking job markets. While DNA offers its own set of advantages, it clearly isn't applicable in every case.

Detective Stan Lipski can't count the crime scenes he's visited during his twenty-seven years as a cop in California. "All kinds of scenes, major crimes, Peeping Toms at dorm windows, even dog nappings! But even with hindsight, even knowing that we likely could have solved more crimes if we'd had DNA evidence available to us, prints are more universally useful. Criminals don't leave prints or DNA deliberately—that's the important thing to realize. But, if you play the percentages, they're more likely to accidentally touch something at the scene than they are to bleed or sneeze or ejaculate over the same item."

Lipski's partner, Catherine Cleary, has only been on the force for six years, and she's a definite fan of the emerging science.

But, like Stan, she's not ready to throw away prints. "Even if we had tons of DNA evidence available at a scene, prints are faster and, if you've got a fugitive situation or a serial offender, knowing who you're looking for is the key."

Lawyers like Judy White prefer print evidence, too. "Juries still distrust DNA. You have to work harder to make a case on DNA evidence, whereas, with prints, you see them start nodding as soon as that examiner says, 'The prints are identical.' There's no waffling, no contamination issues. It's a match or it's not."

Print examiners make one more comparison between fingerprinting and DNA. "If you want absolute individualization, identical twins have identical DNA, but identical twins have *different* fingerprints," White says.

Impression Evidence

THE DISAPPEARING SHOE PRINT

Every day, new technology affects old methods. A convenience store murder/robbery in Toronto, Canada, gives us a look at how digital imaging is taking footprint capture further.

During the robbery phase of the crime, two suspects jumped the counter, leaving a latent print on the surface. Crime scene personnel removed the entire countertop and brought it to the lab. There, one image was taken of the untouched countertop and another was taken of the cleaned countertop, which wasn't moved relative to the camera during cleaning. When the images were digitized and the matching data points eliminated, investigators could clearly see the print of a size 12 Converse sneaker.

While techniques that essentially destroy the original evidence are certainly the exception to the rule, there appears to have been no appeal on evidence presented this way in this case.

Like fingerprints, impression evidence comes in two-dimensional and three-dimensional forms. Shoes leave dirty outlines—two-dimensional images—on a wooden floor or a face; the same footwear also leaves deep impressions in snow—three-dimensional prints. Perhaps because recovering shoe prints shares much in common with routine fingerprint collection, the same examiners often work with both types of evidence. While there are certainly similarities between casting a fingerprint found in a pool of dried blood and a tire track pressed into firm mud, impression evidence presents challenges of its own.

The difference between latent footprints and latent shoe prints is simple. Latent footprints, like fingerprints, contain organic substances: the amino and fatty acids, proteins, and oils that make up normal perspiration. Prints from shoes clearly do not. So, while some fingerprinting techniques transfer easily, others will only muck up the evidence and clarify nothing.

Latent footprints turn up in strange places. One that was discovered on a TV screen proved useful to Paul Gaetan, a Canadian investigator. "Feet typically stay on the floor, but walls, countertops at store robberies, insides of car trunks, and people can all be likely sites. The one on the TV screen in Montreal was odd, but an ear witness had heard the suspect kick something that moved, and we just checked everything we could think of."

Curved screens don't make for perfect prints, but it was enough. "In most cases, floors will still be your best bet—and don't be put off if other people have been in the scene. Good latent shoe prints have been lifted even from scenes where first responders, medical personnel, and witnesses have been through. Of course, if the first people on the scene can identify likely points of entry and protect them somehow until the examiners get there, that's a bonus."

When footwear prints were first collected, an officer dropped onion skin or tracing paper flat over the marks and laboriously traced them with pen and ink or charcoal pencil. That was state-of-the-art criminalistics for the time.

Good light is the key to most forensic investigations, impression marks included. Working inward from any possible point of entry, and not forgetting that window ledges are also possibilities, the examiner first conducts a visual inspection with the naked eye, using white light and alternate lighting sources. Once a likely location is found, ordinary powder (black if the background is light, gray if the background is dark) is lightly applied. Photos at every step of the process are a must. Latent evidence can be lost in collection, leaving the examiner with nothing. Photography also documents orientation, distance between prints in a series, and distance between prints and other objects. Additionally, pictures showing scale must be included in the series, so one-to-one scale blowups can be produced.

> There is no branch of detective science that is so important and so much neglected as the art of tracing footsteps.
> SHERLOCK HOLMES, *A Study in Scarlet*

A Trail of Marks

A single print can be matched to a single piece of footwear, but a series of prints provides other types of information, such as the following:

- The distance between prints indicates the height of the individual.
- The position of feet on a floor indicate how an individual moved. Just the fore part of the foot? A whole print? A print shifted to one side or the other? Depending on whether the person is running or standing or creeping, different parts of the foot strike the surface in different ways. An examiner can determine where a person paused in flight, or

ran through well-lit areas, or stopped entirely, and that information may reveal something about the environment of the crime.

- A trail of prints, even lousy prints, can lead to secondary crime scenes, to an unknown exit, to another set of prints, or to tire impressions connected to the scene.

With a fully developed print, a decision as to what lifting medium to use must be made. The typical two-inch tape used for fingerprints is a possibility, but laying it down in overlapping layers is time consuming and extra attention must be paid to the lines inevitably caused by the overlap. Still, it's a sound procedure that leads to a lift with nothing more than the usual materials found in a fingerprint kit. Other options include the wider lifting material used for palm lifts. Wider tape means fewer overlaps. Commercial applications designed specifically for lifting footwear prints are also available, but wider material has its own problems. It can be cumbersome, and can be more likely to result in air pockets and wrinkles. Rubber lifters, despite the reversal of the image, conform well to uneven surfaces and come in wider material. Practice determines what works best for individual examiners.

Patent prints resulting from tracked mud, oil, or other substances must also be photographed, but their collection requires extra thought. "My first shoe impression was a disaster," Paul Gaetan recalls. "I wasn't a print person, just winging it, but there was no one else available and I figured something—no matter how bad—was better than nothing. Even a shoe size would help, right? The print was an oily substance, but I didn't know what. I waited for it to dry a bit, then blew black powder over it and blew away the excess. The lifting material was overlapping tape that I affixed to a piece of white cardstock and shipped off to the lab down south, thinking 'That wasn't so hard.' "

Even a fifty-seven-year-old can blush.

"The guy who opened it had me on the phone in no time,

gave me an earful on my 'technique.' The 'oil' wasn't oil, exactly. It was brake fluid or something. It melted the tape and we got nothing but a soggy mess for our trouble. Thank God for photos. I think he'd have flown up and shaken me if I hadn't taken photos!"

Faced with the same scenario today?

"I'd tear up the floor and send the whole thing!"

Three-dimensional impressions are the exception rather than the rule in fingerprint work. In other impression work, however, casting or molding is routine. Impressed footprints, tire tracks, and tool marks—the most often cast items—frequently occur outside, making environmental factors immediate concerns. Wind, water, and sun all degrade both footwear and tire tracks. Heavy rain ruins impressions in minutes. All of this explains why some "fingerprint people" travel with huge tarpaulins, big buckets, and even hair spray! (A cloud of hair spray settling over loose material like flour or fine sand can help stabilize the impression. Don't spray *into* the print—just let a cloud settle.)

Good footwear impressions offer advantages over lifts and photographs. Depending on print depth, molds capture details from the edges of objects, and not just from the portions of the bottom in contact with a flat surface. More surface, more opportunity for comparison—and then other advantages appear at court. People respond better to objects than pictures or lifts. For better or worse, one item impressed the jury at the O. J. Simpson trial. It wasn't the minutiae of friction ridges or the cutting-edge technology of DNA identification. It was a glove—a glove they believed couldn't fit the defendant. Being able to match a shoe, a chisel, or a tire to its cast resonates with jurists. On the technical side, there's no question about scale, or fudged photographs, or biased interpretation of any kind.

The fact that a single cast can frequently be individuated to a single shoe or tool or tire makes it important evidence in any

case, not just jury trials. Getting a good cast, then, is well worth the effort.

Plaster of Paris—once the casting medium of choice—has been displaced by dental stone, which preserves finer detail, doesn't require straw or some other reinforcing material to strengthen it, and is tough enough to withstand cleaning back at the lab without loss of sharp detail. The methodology, however, remains much the same as it was with plaster of Paris.

In loose material, or in the presence of water, a casting frame is usually erected. This isn't complicated. A box without top or bottom, or any four flat pieces of wood or other material that can be set firmly on edge around the impression, will do. It should be somewhat bigger than the print so all details are caught. "A springform cake pan does wonders in a pinch," notes Gaetan, who has managed to learn a thing or two in the twenty-one years since he sent a pile of melted sticky tape to a lab supervisor.

There are two basic ways of putting the casting material in the frame or impression. The first is to mix the dental stone in a bowl and drizzle it into the impression with a spoon or stick, starting at the outer edges and moving toward the center. The second is to mix the compound in a bag, snip the corner off the bag, and pipe it in like icing. Either method works. The setting time varies with the amount of casting material. If it feels tacky, it's not ready. When it stops feeling sticky, you can, very gently, lift it from the surface. Starting the lift several inches back from the edge of the form reduces the chances of cracking the casting.

It's not necessary to clean casts at the scene; they will harden further on their way to the lab, making cleaning less risky. What *is* necessary is continuing the documentation process that began with photographing the impressions before casting. While the molding compound is tacky, it's possible to mark on the top surface of the mold with some pointy object. If that's not an option, a black marker can be used to write directly on the mold when it hardens.

Impressions with standing water in them present special problems, as do tracks in snow, but both can be overcome with careful handling. Siphoning or blotting up water can disturb impressions, but actual debris floating on the water, like leaves, should be removed. They could cover part of the impression and ruin the cast. With a clear field, erect the form and begin sifting dry compound over the entire area, including the water. Depending on the depth of the impression, somewhere between a half inch and an inch should be sufficient. While that settles and incorporates the existing water, create a thicker-than-usual compound mixture and spoon or pipe it over the molding, letting it settle and continuing until the entire impression is filled. The "tacky" test determines when these casts can be lifted, but let them sit for at least an hour. Cover if necessary.

PETER HOEG'S SENSE OF SNOW

Smilla's Sense of Snow, a novel by Danish writer Peter Hoeg, opens with a child falling to his death from a snow-covered roof. The title character, aware of the boy's fear of heights, explores the roof in search of a reason for the tragedy. All she finds is a single trail of footprints in the snow. But they're distinctive and will have much to tell:

He wore sneakers, even in winter, and those are his footprints, the worn-down sole of his basketball shoes with the barely visible outline of concentric circles in front of the arch.

When Smilla, a Greenlander and a serious student of snow, attempts to explain to the official investigator what's wrong with the prints, he refuses to take her seriously, though she's certainly speaking his language:

"They were acceleration tracks. When you take off from snow or ice, a pronation occurs in the ankle joint.... If the movement is too fast, not firm enough, there will be a little slip backward.... When you're used to playing in snow, you don't leave that kind of track."

A commercial product called Snow Print Wax stabilizes snow prints, allowing the usual dental-stone casting material to be used even on this difficult surface. Examiners who don't happen to have wax available have substituted baby powder with reasonable results, but neither will prevent snow from melting if warm water—which is usually recommended for the dental stone compound—is used for this application. On top of the requisite steady hand, this process requires cold water—which, unfortunately, also means more effort is necessary to mix the compound and setting times are generally longer. If you use bag mixing, knead the compound well and cut a smaller than usual hole in the corner of the bag. While temperature is obviously a concern when dealing with snow, any such issues are more likely to be connected to the casting materials than to the snow itself. Though obviously designed for use outdoors, Snow Print Wax must be at room temperature before it can be successfully sprayed without clumping or clogging. Complicating the process even further is the inescapable fact that dental stone gives off heat as it sets.

A rather different approach would be to use flour sulphur, which is boiled to produce a hot casting compound. This probably appears like it would be completely counterproductive, but it works very well. Melted in hot water, the sulphur immediately resolidifies on contact with the cold snow and captures good detail. Snow prints, intimidating in their fragility, hold exquisite detail, better than most sands and coarse soils. Generally speaking, this makes them well worth the extra effort.

The process of documenting tire marks should automatically include the search for opposite-side tire impressions, as well. Even if the actual impressions of the tires don't offer any opportunity for photographs or casts, the distance between the im-

pressions provides valuable information about the make of the car to which the tires were attached. The same, of course, is true for front vs. rear tires. Even slips and poor impressions can help determine mechanical details about the vehicle, such as whether it was four-wheel, two-wheel, or all-wheel drive. Investigators may even be able to determine the speed it was traveling, the wheel-base length, and many other factors.

Individualization is the key word in forensic work. Everyone wants evidence that points to one—and only one—item, suspect, or victim. But impressions from tires and shoes, which can be highly individual, also tend to open up more general lines of inquiry. Patterns on shoe soles and tires identify manufacturers, brands, and sizes. And, as databases exist for tires, boot soles, screws, paper types, ink, and hundreds of other items, investigators can not only narrow down characteristics to specific brands, but they can also discover just how many items of a particular type were produced, where they were commonly available, and how long they were in production. Exculpatory evidence (evidence that proves an individual or item is *not* the one sought) saves investigators time and effort, so knowing the number of possibilities makes it easier to assign the right number of investigators to promising leads and to eliminate suspects more quickly.

Tool marks, which appear most often on vertical surfaces like doorframes, obviously can't be cast by pouring compound into them. On the upside, though, doorframes and other surfaces holding tool marks generally don't melt or get washed away. They're usually visible and, if necessary, can often be physically removed and taken back to the lab with little risk of transport damage. Several different casting materials, usually with a rubber/silicone base, are commercially available and easy to use. Mikrosil is commonly used and, like epoxy, comes as two different compounds that are mixed together at the scene and then pressed into the tool mark. These casts are often called "popsicles" because popsicle sticks are used to mix and apply the casting material, and the sticks can be left in the cast while it hardens, providing an instant handle. These sticks serve another purpose, as well. Mikrosil won't take a mark once hardened, so notes can be jotted on the stick. While not particularly delicate, they can be damaged by pressure, so they should not be stacked.

Examiners classify tool marks into two types: friction marks (formed when some part of the tool rubs against a surface) and stamping marks (which show a full impression of the edge or face of a tool at rest). Many impressions are combinations. A chisel traveling across a wood or metal surface creates streaks that, at high magnification, can be seen as very distinct stripes. Where the tool comes to rest, a stamping mark of the leading edge of the tool is created. Both types should be photographed and cast. If the tool itself is available, comparison should be made by photograph or cast. Actually attempting to fit tools to marks usually ruins the marks, even if the right tool is found. Tools found at the scene should be bagged so that any trace evidence clinging to the working surfaces is preserved. Having samples of the actual worked surface makes other types of

chemical and microscopic matching comparisons possible, so these should be collected and brought to the lab whenever it's practical.

LUCAS WILLIAMS

In the middle of an open homicide case, Lucas Williams has lined up nearly forty of the sharpest looking saws, blades, and industrial cutters that he could find. He is slowly sawing through pig femurs with each of them. The stench of burning bone is overpowering.

"I really hate dismemberments, but they do usually provide us with good toolmark and match evidence." A high-pitched whine raises goosebumps as a reciprocating saw dances briefly on the bone before biting deep. "Even if we don't find the actual tool, we often determine what sort of implement was used, and that can be very useful." Peering closely at the cut end, he shakes his head. "Definitely not."

Before he can pick up any other pointy objects, Lumi Obwani sticks her head in the door and announces that they've found another piece and the medical examiner will let them know when they can collect photos. It's hard to imagine an occupation where finding more body parts is a *good* thing, but Lumi's parting line—"Every bit helps!"—sounds distinctly cheerful.

Grinning at his interviewer's obvious discomfort, Lucas deliberately revs the next small garden appliance a little more than usual before tackling another section of pig. "Too bad you didn't come on a day when we were looking at a stabbing. Much quieter." Another cut, another shake of the head.

"In this line, you don't really get 'odd' cases—just cases with more or less common implements. But, in that vein, Lumi and I did learn a lot about Japanese sword making a few years back."

Leaving his grisly task for a moment, he carefully washes his hands before heading into his office. He pulls out a beautifully illustrated book featuring hundreds of gleaming blades superimposed over images of men in what looks like straw armor. "Neat, isn't it? Perhaps the finest metalwork in the world being carried by men wearing next to no defensive gear." Flipping ahead a few pages, he points to an image of two men leaning over a tiny forge. "Modern-day work that purports to be in the old tradition. But, of course, it doesn't even come close—at least not in the ways that count to criminalists." Pulling a tiny capped test tube from a shelf, he points to the edge. "Most of the so-called swords seen here are blind—they don't actually have a properly sharpened edge. They also don't have the impurities found in older blades. That bit there, though—that bit was the real McCoy. It was taken out of a victim found dumped up in the valley and, when I first saw it, I wasn't really sure what I was looking at. Under the scope, it had a different interior consistency than most things I'd seen. When we did the chemical analysis, it turned up really odd combinations. While we hoped to get simple match evidence if the investigation eventually produced a weapon, we also thought having a manufacturer or country of origin might reveal something useable."

Lumi sails in through the door with three steaming cups of coffee and, on seeing the tube with its little fragment of metal, nods. "I figured he'd pull out that one. He tell you about the fun we had figuring out where it came from?"

"Just getting there."

Nabbing a book that could just as easily function as a small coffee table, Lumi flips rapidly to the back to display page after page of tables. "Exemplars of thousands of metals, everything from elegant dinnerware to dental fillings to cast-iron frying pans." With one finger, she flips it closed again. "And that thing didn't match any of them, not even close."

She curls up on a seat. "Still, there's only so many things that usually get jabbed into people and we were reasonably sure, from the depth of the wound, the angle, and the damage reported by the M.E. (medical examiner), that it was a long knife of some sort."

"So, we went to the museum and started eyeing all *their* sharp, pointy objects."

"We knew it was a single-sided something, which helped—"

"But not as much as the Chinese food and the Jackie Chan movie Lumi dragged us off to after we finished at the museum."

Lumi smiles, shrugs, and sips, letting Lucas describe the evening.

"We were pretty well primed from having seen all those old swords and daggers and stuff, but when this fellow jumped out of the bushes and started swinging around this blade, it occurred to us at the same moment why we couldn't find any modern match for the metals. It wasn't modern. It was old. Really old. We never did see the rest of the film—just rushed out and started looking for an expert in old Asian swords."

"Oddly enough, they don't have a Yellow Pages listing for those."

"But, once we did find one, he helped narrow the weapon and time period, then provided the names of the dozen or so people who were most likely to know something about who, outside the museum community, was both local and in possession of such items."

Lumi, her coffee already gone while Lucas's grows cold, tips her head toward the ceiling, above which are the offices for the department's detectives. "I think they thought we were joking—called it the "Highlander Case" when we told them—but, eventually, they found the weapon and the perp." Swinging out through the door again, she calls back, "Pity he didn't look anything like Adrian Paul, though!"

"No, he definitely didn't," Lucas agrees. "But we did get a solid ID on the weapon with the usual suspects in toolmark work—match evidence along both the fracture between the blade and the chip and between the blade and the marks on the victim's bone—as well as the chemical match between the blade and this little fella." He shakes the tube and tucks it back on the shelf. "There are two lessons to take away from that, of course: one, don't eliminate any possibility regardless of how remote; and, conversely, two, don't go looking for authentic antique Japanese swords in every kitchen drawer, either."

With that, he's back to the pig bones and Black & Decker's latest entry in the low-end home saw market.

The whole idea behind impression comparison is that both the impression *and* the object can be collected and examined for common distinctive details. Occasionally, it proves necessary to "print" the item itself, especially if the object is suspect but there is nothing to match it to and there's no apparent reason to hold the original piece as evidence. Tools, tires, and shoe soles can all be photographed, but natural pressure distorts some items—tires more often than tools—and it may be preferable to capture an image of the marks the item would normally leave instead of a picture of the item itself.

Printing a shoe is simple—roll it in ink and step on a piece of paper. But printing a whole tire takes coordination. Running a tire over paper affixed to a subsurface usually destroys the paper, tearing and wrinkling it. So, after the tire is inked with a large roller, the car should be slowly driven over unsecured paper, which should be removed before the rear tire can smudge the prints. If the wheel base is short, several sections of tire can be printed and the resulting sheets of paper can be stitched together.

Tools, which have little sensitivity to pressure and can't be printed in this fashion, can still be photographed and molded.

The last type of impression prints commonly faced by examiners are dust impressions. Like dust fingerprints, these can't be powdered and lifted. The brushing would destroy the layer of dust. Photographing dust is extremely difficult, even with adequate lighting. For these impressions, a mechanical tool—the electrostatic dust lifter—can save the previously unsalvageable.

Silver lifting film, large enough to cover an entire shoe print without seams, can carry an electric charge and, after being zapped, attracts all the dust and debris directly beneath it. Some surfaces yield better results than others, and the dust impression isn't always visible on the film, but, as residual charge can attract dust, every film should be covered as quickly as possible and turned over to lab personnel. Procedures that can't be directly applied to the dust layer on floors or tar roofs can be applied to the dust layer on the lifting film. Electrostatic lifts, though not useful for collecting a print, can clean up a dusty surface before gels or other techniques are employed.

Kari Day-Wells, who inherited the job of electrostatic dust

lifter by default, nearly threw away the evidence after her first try with the new gadget back in 1983. "I couldn't believe it. The department put out a significant amount of the budget for this thing. The company sent in a trainer, but, of course, the person who was trained with it promptly went to some other jurisdiction. I don't know what I was expecting, but it was certainly more than I saw that day. I've since discovered, of course, that some results are spectacular, absolute magic, but that particular lift looked like nothing. Two days later, when that lift went to Malcolm Cole in the next town over, I took it myself, thinking I'd messed up somewhere. I was willing to beg a crash course. What Cole brought up out of that nothing was this beautiful, big, sharp, clear print! You could see tiny cracks and individual fibers in the stitching on the soles. I wanted to take a picture of it and frame it, it looked that good. And I almost threw it away."

Hair, Trace, and Fiber Evidence: Found Treasures

"The dark hole of envelopes, that's me!"

Jay Paresh often describes the flurry of evidence packets arriving from scenes as "Cracker Jack prizes." "Trace evidence covers everything from tire dirt to broken glass to carpet fluff. We turn organics—that would be seeds or leaves . . . ah . . . monkey hair we had once . . . basically anything from living sources that aren't human—over to our forensic botanists and zoologists, and most human trace except hair, like blood or semen stains, over to serologists. But, everything else comes to us at some point."

Which leaves plenty for Paresh and his colleagues to play with.

"I don't recommend investigators vacuum whole scenes. It's just not a good use of their time. But good investigators can analyze a scene and make informed choices about promising locations. Most are excellent observers, picking up on anything out of place or that might contain more evidence, like a cigarette butt. They're intuitive, but it's an educated intuition. Small departments faced with a major case tend to send everything. Which isn't a bad thing, just time-consuming. Experienced criminalists send us less stuff, but it's generally more meaningful stuff."

And meaningful stuff would be?

"Dirt is typical from most scenes. The criminalist concentrates on areas the suspect *had* to be in contact with—anything directly in front of an open store safe, for example. They'll ignore potting soil on a window ledge if the window is painted shut and there are herb pots there. Common sense tells them that the potting soil didn't come from a perp sneaking in through a window. No one wants to leave evidence at a scene, but no one wants to clog up the system with irrelevant material if they can help it, either."

Trace examiners see lots of dirt, and many departments hold geology reports on their jurisdiction in the hope that an unusual sample will lead to specific locations. "Even relatively consistent areas show some variation. Take the Vegas area as an example. When dirt from a tire tread showed lots of organic material—leaf mold, mossy bits, and mud—we started looking at Mount Charleston. It's one of the few places where those conditions *could* exist."

Exemplars (samples taken from known locations) line the shelves of Jay's garage. "Lincoln Rhyme is my dude. My son thought I was the dullest guy on the planet until he saw *The Bone Collector*." It might take a writer of Jeffrey Deaver's caliber to make dirt interesting to the layperson, but for trace examiners it offers plenty of intriguing information. They'll crawl into the strangest places to get it. "If you hear a Dustbuster and see nothing but legs sticking out of a trunk, or some guy picking grains of sand out of floorboards with tweezers, it's probably a trace examiner looking for dirt. They're also the ones with little bottles in their glove compartments, the ones with bottles

rattling in their pockets. Compulsive collectors. A vacuuming gets lots of other things, but dirt is a primary target."

BONIA CAPPITELLA

"Years ago, before I ever thought of forensics as a potential career, I was watching an episode of a rather horrible mystery series called *Burke's Law* when the scriptwriter decided to have the killer stab the victim with an icicle to the heart. There was a whole bizarre setup with an ice sculpture shaped like Cupid and the ice arrow was rigged to fire remotely or something and, of course, the arrow melted and left the investigators without their usual clue, the murder weapon.

"Even then—and I wasn't very old at the time—I remember thinking to myself that no one would go to that much effort, that much *unnecessary* effort especially, just to ensure the cops wouldn't get their hands on a murder weapon." Bonia Cappitella's laugh is partly rueful, and partly delighted. "But, I should have remembered my mother's favorite comment during the evening news: 'Look, see? Truth *is* stranger than fiction!'"

Not that Bonia ever had to investigate a cherub's alibi, but like most investigators with a few years under her belt, she's certainly encountered weird situations.

"Well, the one that caught my imagination most didn't even involve me. But it was so convoluted, I've just never forgotten it." Bonia's supervisor was often sent to one of the remote fly-in-only communities in Canada, and the story he always used to illustrate the need to observe everything revolved around a chunk of wood.

"Local police were called to a small cabin early one morning where they found one man dead, another pretty woozy, and a distraught widow who swore spirits must have killed her husband. Frankly, any investigator would have had to look twice at this one. The wife arrived home from town that morning because a heavy snowfall had made it too difficult to find her way the previous night. The two men, her husband and a visiting trapper, had clearly not left the cabin since the snowfall ended late that night—no tracks—and, equally obviously, no one had arrived, either. Yet the doctor figured death occurred about four in the morning, well after that. Two victims, one with his skull smashed in, another lucky to have avoided the same fate—his injuries were, likewise, blows to the head with some blunt instrument—and, on first investigation, no assailant and no weapon inside what was a pretty small dwelling."

Bonia's grin flashes again. "The classic locked-room scenario!"

Despite close questioning, the trapper claimed to have no memory of the events of the previous evening. Given the severity of his injuries, medical personnel were willing to concede that it was entirely likely that traumatic amnesia was at work.

"Still, my super, he wasn't the sort to believe spirits could whale the tar out of two perfectly healthy men, so he was all over that cabin and the surrounding area. The one thing he did concede early on was that no one had come into or left the cabin. Far from confirming the presence of ghosts, that only convinced him that the surviving man wasn't a victim, but the murderer. The only question was how did it happen and where the heck was the murder weapon."

Not an icicle!

"No, but close! Like most cabins, this one was heated by a pair of wood stoves. The first officer on the scene had noted at the time that the stove in the bedroom was cold, but that there was still significant heat in the main room and that the coals of the previous night hadn't burned down completely. The M.E. had to know that as well, to try and calculate a time of death, but no one really paid much attention to it until my super suggested the trapper and the cabin owner's injuries were completely consistent with two men beaten with a chunk of firewood."

Which was easily tossed into the fire!

"Exactly! We'll never know for sure how the whole thing started, but it seems likely that an argument between the men turned violent, that one of them, probably the cabin owner, struck the visitor several times with the nearest bat-size item—a piece of firewood—and then was either attacked himself with yet another piece of wood or, perhaps more likely, had his weapon taken from him and the tables turned. In either situation, the trapper must have realized that the improvised bat was evidence of some kind, so he simply stoked up the stove and tossed it in. Which also explains why it was still so warm in the cabin when the wife arrived home."

So, not quite the melting icicle of the *Burke's Law* scripts, but close.

"Mmmm. As it turns out, I was the one to run into an icicle as a potential murder weapon, and it happened in the Greater Toronto area, not some little village up north.

"It was my first winter in Toronto, and I'd seen pretty routine work up until then. It must have been about two-thirty in the morning when I got the call out and, when I arrived, it was still a pretty chaotic scene. Typical domestic case in that sense. Kids screaming, neighbors wandering about in nightclothes trying to give statements and rubberneck a bit while the girlfriend wailed at the top of her lungs and the medical examiner hovered over the victim—a Haitian man of about thirty who was lying on his front step with his head cracked open."

No firewood in sight?

"No, not even a twig to be seen."

Neighbors had reported an intense argument to police nearly an hour earlier but, when they'd checked, there'd been no evidence of violent activity between the two. There being no history of violence at the address, the police had left with just a warning to keep it down.

All had indeed been quiet in the intervening hour but, according to the oldest child, a boy of thirteen, the argument had ended with his mother insisting the boyfriend get out, that she didn't want him around any more, and that she'd call the police back if he didn't leave.

Not all the neighbors had gotten back to sleep by that point, and at least two of them had heard raised voices again and peeked through their curtains long enough to see the boyfriend standing on the step, banging on the front door, before going to call the police once more. No one actually saw the girlfriend open the door just moments later, but they all heard her screaming and arrived to find the boyfriend lying on the step with blood pooling around his head. Her insistence that she'd just come to the door to toss his car keys out was met with the neighbors' and investigators' obvious question: If the girlfriend didn't hit him, who did?

"There'd been sporadic observation of the house throughout and there was no one else noted in the vicinity. And, even if there had been, it would have been an extreme coincidence for someone who wanted to kill her boyfriend to arrive on her step just as she was throwing him out of the house. It really looked as if she were the only reasonable suspect."

Yet she was never charged?

"Well, despite the sort of stereotypical run of events, there were some things that just didn't make sense to the officers first on the scene. Or to me. First of all, why kill him on the step, in full public view, if she hadn't done it inside? It didn't make sense to assault him, with or without the intention of killing him, if he was already out of the house, which he certainly was at that point. And there was the original report from just an hour before that the argument, though certainly loud and heated, hadn't shown any inclination to violence."

Escalation?

"If the man had died inside the house, I think we'd have given that possibility more credence, but he didn't. And the children weren't that young—thirteen, eleven, and nine. They were credible witnesses, and they reported that they'd heard him demanding his car keys and that their mother had spent the time while he was banging on the door rummaging through her handbag for the keys. When officers arrived, after a very short response time, she still had the keys in her hands and her purse was upended in the middle of the kitchen table. All the little things were supporting the statements from both the woman and the children—except the boyfriend was dead.

"The M.E. was in the process of moving the body when I came back outside and, as it was coming on dawn by then, it was the first chance I'd had to really view the entire scene without the screaming and distractions of the extraneous people around. As I was coming down the steps, I noticed dozens of little ice chips on the treads. In fact, one of the officers warned me to watch where I stepped or I'd probably end up falling. That's when I realized what had likely happened and looked up. Sure enough, all along the eave were icicles, the monster ones you get in Toronto, where the temperatures during the winter rise and fall around the freezing point day in and day out for weeks at a time, and, dead center over the front door was a clear space with not an icicle to be seen."

The melting weapon.

"The boyfriend hadn't just been knocking on that door, he'd been pounding on it. The vibration loosened the ice above him. It fell, striking him in the head, and, as he fell from the first blow, he struck the other side of his head on the concrete step. At autopsy, the M.E. found sufficient injury from just the falling ice to have killed him, even if he'd landed in a feather bed afterward. So, technically, the weapon—if you could call it that in this case—really was an icicle."

Hazardous Evidence

Investigators and examiners live by the maxim "Whatever you don't expect to see is precisely what you'll find." Consequently, they have to be prepared for hazards at any scene.

- **Chemical threats:** Any compound injurious if ingested, inhaled, contacted, or ignited. Drug labs are primary threats, but others exist. Bomb factories, bomb scenes, and the scenes of industrial accidents pose serious chemical threat. Respirators and other personal protective gear are necessary at such sites. For the safety of others, items in custody that are suspect or are known to be dangerous should be appropriately labeled by the collector.
- **Biohazards:** Blood and other biological fluids can carry hepatitis, AIDS, or other viruses. Trace that has been in contact with them should be marked "BIOHAZARD."
- **Explosive threats:** Gunpowder, blasting caps, and cord may all indicate the presence of explosives. The possibility of secondary material that is arranged—whether accidentally

or deliberately—to harm investigators should not be ignored. Materials from these sites should be marked as potentially explosive.

A few of the other items trace examiners look for include hair (human and otherwise), glass, lost items like earrings or false nails, and fibers of all types.

"We also collect 'match' evidence," adds Paresh. "In kidnappings, we'll collect any tape or rope at the scene. If we recover other evidence elsewhere, it might be possible to match two ends of a roll of tape, proving a connection between the first crime scene and the new location."

> He explained that when he had used hot blowing air to open the tape receipted to him by the Black Mountain Police, he counted seventeen pieces ranging from eight to nineteen inches in length. Mounting them on sheets of thick, transparent vinyl, he numbered the segments two different ways—to show the sequence the tape had been torn from the roll and the sequence the assailant had used when he taped his victims.
>
> *The Body Farm*, Patricia Cornwell

The obvious purpose of all these efforts is to resolve a crime and convict a perpetrator. But in some cases, match evidence functions as identification evidence. "Match evidence on a roll of tape was the only way to identify a body found in northern California when I was a trainee. The body was in an advanced state of decay—no prints, no identifying clothing, nothing specific. 'Three-feet something, blond, female, no old bone breaks,

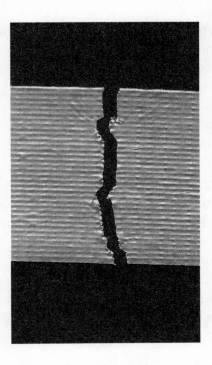

about seven years old' just wasn't enough to eliminate or include specific people. The duct tape still wrapped around her wrists—*that* matched the roll collected from her home over a year before. It wasn't the outcome we wanted, but her family was grateful just to know."

Trace evidence examiners can work as generalists at the scene, collecting trace as well as tool marks, fingerprints, and other material, or as specialists. "Glass is one of my interests, and I've made it an area of study. I can do a lot with modern glass headlights, for example," says Paresh. "But there are people who can identify formulations in eighteenth-century Venetian glass, so clearly they're working on a whole different level."

In break and entries and some other major cases, the physical properties of broken glass are the real issue, not the source. Working from even tiny fragments, a good glass expert can determine whether an impact came from inside or outside a building or car, the prescription of broken eyeglasses, or whether two pieces of glass might have come from the same pane. "Fractile characteristics aren't in the crime scene investigator's head, but engineers and optical physicists create tables that examiners can refer to, or they publish papers that at least provide us with guidelines and contacts for that sort of expertise.

"Automobile glass, because you find it everywhere at accidents and hit-and-runs, is well-documented. In some cases, a piece of auto glass can narrow the possible vehicle to just one or two makes."

Paresh recalls a case in Iowa where a man walked into the station because of an aired report that police were looking for a Toyota Corolla. "We made the ID based on glass, but he assumed there must be a witness and turned himself in."

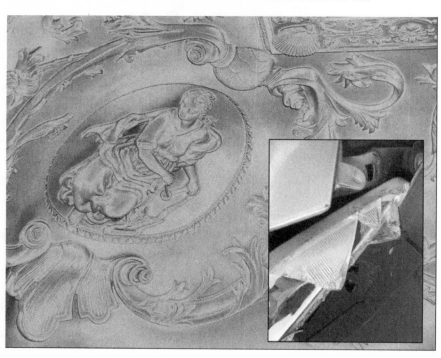

> I could see where a bullet had exited through the window directly behind Cross's bed. The glass fracture was clean and the radial lines even: The shooter had fired from a standing position, directly across the bed.
>
> THOMAS PIERCE, FROM JAMES PATTERSON'S *Cat & Mouse*

Glass, with its fractured edges, provides opportunities for edge matching, making it imperative to collect as many pieces from the scene as possible. Like a huge jigsaw, examiners can slowly reassemble the bits and pieces. Cracked or holed glass tells examiners whether damage resulted by low-velocity assaults, like a fist or rock, or high-velocity assaults, like bullets or explosives.

Metal shavings, usually the result of tool work, are routinely collected at the crime scene. Most derive from activities undertaken on the scene and can be quickly attributed to locks or cash boxes. The action of prying open metal doors often results in curls of metal from the door or its frame, which are found on or around the lintel, but the process can also cause metal to be shaved from the tool used. As is true with all trace evidence, the hope is to eventually match shavings to items brought to or taken away from the scene. If the metal's composition is rare or unusual, shavings can help identify the tool itself, sometimes down to an actual brand of tool. At that point, investigators can start searching for sales of similar equipment.

Something else that can get transferred during forced entries or vehicle impacts is paint. Paint is the favorite evidence of many trace people. "It gives you so many opportunities," Paresh says, holding up a chip as big as his nail. "First, it's match evidence. Irregularities on the edge can be matched to a tool or a car or painted furniture. Auto paint is generally applied in layers, so the physical and chemical structure of those layers provides several clues. Then there are visual properties like color matching. On cars in particular, colors can be specific to a particular make, model, and year. That's huge."

Once the chip has been photographed and color matched, it can still reveal information during microtrace examination. "Microtrace examination includes chemical testing that can destroy a sample. So it's the end of the analysis process, but it can give us specific formulations that, when compared to known exemplars, can pin down a particular type of paint or manufacturer and we can then trace the product from factory to seller and, hopefully, to buyer. Or another sample."

Paul Gaetan, who learned all about tape and chemical reactions with that first shoe print lift, adds, "But don't pick up paint chips with tape! That can ruin the microtrace results. Get out the tweezers or a brush and get the pieces into an envelope, but don't pick them up with tape."

Trace Examiner's Field Kit

Keeping in mind that all crime scene responders carry a large amount of equipment, and that many wear several different hats at the scene, a field kit for trace evidence should contain the following:

- evidence vacuum, extra batteries or power cells, extra disposable filters
- photography equipment
- glass slides, slide covers
- tweezers, chopsticks, other manipulating equipment
- light sources—regular flashlights, alternate light sources (UV, IR, etc.), lasers—with power supplies
- measuring tools, including protractor
- bindle paper (non-reactive paper)
- lifters, acetate covers

- lifting tape, household tape
- envelopes, bottles, boxes for collection and transport
- cutting implements—saw, knives, scissors, etc.

Back at the lab, the examiner may have access to more powerful tools, such as a stereo microscope, comparison microscope, polarized light microscope, visible microspectrophotometer, and micro-Fourier Transform Infrared Spectrophotometer.

Clothes, blankets, drapes, carpets, auto upholstery, and all fabrics (natural or synthetic) shed fibers. Even the seemingly indestructible polyester releases fragments. Like other small trace evidence, fibers can be manually collected with tweezers or vacuumed into filters. Torn pieces of cloth may constitute match evidence and can be preserved and compared to other samples. Because tape residue contaminates fiber evidence, lifting tape isn't the preferred collection method. Other techniques, such as static lifts (which are covered in non-adhesive acetate) speed up the process and pose no risk.

The vast number of fiber types appears to rule out the possibility of ever getting a match, but both normal and microscopic examinations reveal a great deal to experienced examiners. First, if exemplars of all fabric related to individuals known to have legitimate business at a scene are submitted, useful fibers that might implicate others can be identified by process of elimination. (The exemplars, of course, can also serve as comparisons for fibers on the clothes or fabrics belonging to any eventual suspects.) Next, when one of the scene's suspect fibers is mounted on a scope, details of its construction come into focus. With the simplest lenses, a determination of whether the fabric is natural or synthetic can be made. Examiners can also determine the color, the twist or kink of a fiber, and, in natural fibers, the relative length of the individual strands that make up the fiber.

Alternate light sources illuminate some chemicals applied to a fabric. Stain resistors, fire-retardant sprays, and applied dyes may fluoresce in different wavelengths. Chemicals that the fabric might have contacted previously can appear, as well. "We can burn parts of larger pieces for spectroscopy results," says Donna Simms, a chemist. "In drug-related cases, it's not unusual to test for the drugs themselves or for components used in drug manufacture."

Cordage Characteristics

A rose may be a rose, but rope and cord (and fabrics or fibers) can be described by—and therefore differentiated by—dozens of characteristics. The following is a suggested listing of cordage traits that the FBI considers a "starting point" for describing rope or cordage samples:

- diameter in cross-section
- presence of staple or filament fibers
- twist, braid, or non-twist
- type of twist
- crowns or turns per inch
- number of plies or braids
- twist of each ply or braid
- filaments in each ply or braid

Filaments, the last trait listed above, are further broken down by these:

- core, if any
- twist
- color or colors
- coatings, if any
- tracers

Who knew rope was so complicated? Add in the FBI database of knots in rope, wire, cord, and other bindings, and rope becomes important evidence well worth preserving along with other fabric and fiber evidence.

CHI HU

Chi Hu has twin daughters who love to play cat's cradle, a game in which a string is looped in knots on one player's fingers and transferred to the fingers of another so as to create a different pattern. But Chi herself has a whole different take on knots. She examines them as part of ongoing criminal investigations. It's anything but child's play to her.

"I don't know anyone who studies knots as a full-time occupation, at least not in departments of our size. But over the past few years, databases of knots and ligatures, patterns of confinement of victims, and unusual uses of rope or wire have certainly been recognized as sources of information in their own right. We used to just hope for match evidence on the cut ends or in the material itself, but it's much more complicated than that now."

A typical knot investigation begins on site, and Chi Hu was once called to observe the removal of a hanging victim from a college dormitory.

"The attending officer had nearly twenty-six years experience in the field and, while little of that would have included hanging as a method of murdering someone, there was something about this setup that didn't fit with his mind-set of what a suicide should look like."

John Purdey, the officer who asked Chi to attend the scene, still doesn't have a full picture of what exactly seemed wrong that morning.

"At first, I think my partner and I were both simply startled to realize that the hanging, suicide or not, had been undertaken with the victim's feet still on the floor. There was a case in lockup once that I'd seen myself, where the guy had tied a piece of sheet to the upper bunk and simply lowered himself to a near-sitting position on the floor and asphyxiated without jumping or breaking his neck. But, you know, that's—at least to my mind—a pretty tough thing to do to yourself."

Evidence of other suicides certainly agrees with that assessment. In almost all cases, the suicide-by-hanging steps off something in order to eliminate any possibility of second thoughts. Cases of asphyxiation where the victim could simply have stood up to save himself or herself are, for the most part, associated with autoerotic or unusual partnered sexual practices, and the intent is never to actually die or kill.

"I guess that was what made me look twice at this scene—that there was no evidence of sex play. But it was certainly unusual for someone to have the willpower to hang themselves without the usual step-off."

At the scene, Chi Hu had difficulty seeing the actual ligature and knots. "The body had been there overnight, and swelling around the ropes was impeding view."

The local medical examiners realized that evidence could be destroyed if ropes were simply removed from the corpse, and they had no objection to Chi taking her own images, both video and still, as the body was examined and the cords removed. By cutting through the cords—not the knots—and carefully documenting the process, she could ensure that she would have a clear picture, even months or years later, of every move required to recreate the knot patterns and confinement paths of the ropes.

"The first thing you'd want to know if you're determining whether someone tied themselves up, or if someone else did it for them, is the handedness of the victim. You can't necessarily tell from the ligatures if the person was right- or left-handed, but the chirality—the usual direction of movement—is generally consistent. In this case, I could, as the ropes were removed, see the motions required to get them in that final configuration. And, like John, I started to have serious concerns about whether it would have been possible to complete those knots at all with the victim holding her hands behind her own head! There was no scraping or rubbing under the rope, so we knew that there was no scenario where the knots were tied in front, then twisted around behind."

Chi immediately ensured that the rope—a heavy, twisted twine—would be swabbed for possible DNA evidence, then ushered it through the process of being photographed at high magnification, with particular attention paid to the end cuts so that match evidence remained a possibility. With all the possible trace removed, she spent nearly an hour just staring at the carefully aligned pieces. Then she began tying up her staff.

"Sometimes, it's important to have a firm idea in your mind of what happened. At other times, fixed images can blind you to the possibilities. With knots, you have to be able to visualize each step in their creation clearly, then see if it is remotely possible to complete those motions from various approaches. In this case, I was becoming more and more sure that the victim could not have physically performed the necessary manipulations. And that wasn't the only anomaly. These weren't your garden-variety knots."

In the victim's apartment, Chi looked for evidence of other knots.

"Most people have at least one pair of shoes that they leave tied up and just shove their feet into, but, other than that, we're really lucky if we find anything else. The victim's room was typical, dorm-room small, and we were surprised to find *three* examples of knot work—one set of tennis shoes and two other pieces."

The shoes were typical knots and bows—nothing of note there, except that the victim had added a few little cubes to the laces to spell out her name.

The second item seemed to be a dead end—a macrame plant holder done by a floor-mate. "To the best of our knowledge, the victim herself did no crafts, so we had no samples of her own for comparison."

The last item, a bulletin board with ribbons crisscrossed over it to give it a Victorian appearance, contained a "lover's knot"—two locks of hair about three inches long that were tied on one end, braided together, and tied again on the other. "Not a typical memento, but with the bulletin board and several other objects in her room suggesting a fancy for Victoriana, it wasn't as out of place as it might seem."

Reviewing the three items and comparing the knots, Chi found some knots in the cords that were like those in the macrame work. But, as the knots were common ones, there was no definitive point of comparison. The "lover's knot," on the other hand, was another matter altogether.

"It's rare indeed to find knots that are specific to one sort of occupation turning up in other places. In the knots at either end of the braid were unusual slip knots that held perfectly well when tugged tightly from either end but that, with just a twitch in the opposite direction, would release the entire tangle. The same knot appeared in the noose. We see these knots fairly often in ship's rigging, but how many people are there alive today who use—correctly—old sailing knots?"

Staring at the odd configuration, Chi questioned her hypothesis that this was in fact a staged murder, and not a real suicide attempt.

"With slipknots in the mix, and evidence that the victim was familiar with that type of knot, I had to wonder if we were seeing proof that the victim *thought* she could extricate herself from the noose. In overdose suicides, we often find that the person actually took smaller doses of pills at earlier points in time. In suicides where the victims cut their wrists, you frequently see 'hesitation cuts'—shallow cuts done prior to a cut that's fatally deep. It's like they're working their way up to the actual event. With the victim in this case having her feet still on the floor—so she could have stood up at any time until she actually went unconscious—and the slipknots that could, theoretically at least, have allowed her to abort the hanging, I was wondering if this wasn't another form of suicidal lead-up, but one that had gone bad unexpectedly. It would also explain the lack of a suicide note, if she hadn't actually meant to kill herself **this time**."

Confronted with three possibilities—murder, suicide, or death by accident—Chi went back to the notes taken at the beginning of the case.

"No matter how I looked at it, though, it was obvious that, whatever happened later, this young woman could *not* have made those ligatures herself, not behind her neck. She may have believed she could stop at any point, but she couldn't have started out alone."

"Suicidal individuals often indulge in highly symbolic behavior before death," notes John Purdey. "When Chi came to me with the information she had so far, I was struck by the lover's knot. It was such a romantic-sounding thing, but nowhere in the interviews we'd done with friends or family had we come up with any indication that the victim herself could have made it—that she actually had the knowledge to do so."

Working on the Holmesian theory that removing the impossible leaves only the possible, regardless of how improbable, Chi and John approached the case from a different angle.

"Instead of asking if she could have made the knots herself," John recalls, "we started asking who in her circle could have made them *for* her."

"The obvious person was whoever had made the lover's knot in the first place." Chi shakes her head. "It sounds simple in hindsight, but it wasn't at the time."

"Though," John adds, "there was one other factor working against us. The whole notion of a lover's knot suggests a *lover*, but according to all the people we interviewed, there was no boyfriend, no girlfriend, no lover of any kind."

"Which, in the end, was the key to the whole thing." Chi's fingers make knots of themselves as she thinks back. "With no lover in the picture, the next obvious person to question was the floor-mate, the one who'd created the plant holder. She was the only other person in the frame who could possibly have done it, who might have had the knowledge."

"That was a mess," adds John. "I think, in the end, she just wanted to tell us, wanted it over with."

"Wanted someone to understand."

"We assumed, because a lover's knot seemed so personal, so

intimate, that either the victim or the victim's lover would have made it. When we couldn't find a boyfriend, and the victim didn't seem capable, we thought we'd come to the end of the line there."

"But it wasn't either of the owners of the hair who made it. It was the floor-mate who took what had, until then, been a simple braid and fashioned something more in keeping with both the victim's style and the floor-mate's interests."

"Once we'd established her as the author of the lover's knot, it wasn't a big jump to her as the author of the knots in the ligature, either." John pauses. "I'm still not sure, in my own mind, if she meant the girl to die or not, but she certainly helped in the planning and the execution."

So, was it murder, suicide, or accident?

Chi folds her fingers flat over each other. "She claimed at the inquest that all she did was tie the knot behind the victim's neck, that she knew her friend was depressed over a relationship back home that had come to an end, that she was aware of the importance her friend placed on the lover's knot, that she had indeed fixed it when it had originally fallen apart, but that she'd never thought the victim would actually kill herself. As there was no evidence of a struggle, and there was every indication that the victim herself hadn't intended this to be the moment when she actually did commit suicide, and as the floor-mate had nothing obvious to gain from the woman's death, and, furthermore, as it can be established that, at the time of death, the floor-mate was in the common room downstairs, it was ruled accidental death while engaged in possible pre-suicidal activity."

It sounds like Chi has memorized parts of that speech word for word from the decision paper.

"Whatever that means."

Like Chi, John Purdey still doesn't look convinced. "I just can't think what normal human being would help a friend tie a rope around her neck, then simply walk out of the room!"

Chi shakes her head slowly. "A jealous one."

That, of course, is a purely subjective opinion, one she couldn't include in a simple report on ropes and knots.

More discriminating scopes can examine cross-sections of synthetic fibers. Extruded fibers like nylon and similar products begin life as liquids, then get pushed through a nozzle of some type. Different nozzles have different shapes. Carpet fibers in particular show differences—some look like squares, others are trilobular, or oval, or circular. With a combination of chemistry, shape, and color, it's possible to make confident statements about whether one sample is consistent with another or not. The Atlanta child murder cases generated thousands of exhibits, among them some rather distinctive orange trilobed fibers from carpeting.

Despite the possibility of it providing DNA evidence, hair can be treated as any other fiber. It shares many qualities of wool and, like other fibers, can be classified by color, by twist or kink, or by dyes or other treatments applied to it. It may be possible to determine ethnic origin—though hair examiners point out that, with an increasingly multiracial population, such distinctions aren't always possible. Determining where on the body a particular hair comes from—especially if the hair is entire—is much easier. And deciding whether the hair fell naturally or was plucked is as simple as noting the relative angle of the root to the rest of the hair shaft.

Unlike synthetic fibers, hair—a natural element—picks up characteristics of the body to which it was formerly connected. An array of drugs and other substances are laid down in the shaft of the hair, making it possible to gather information that will or will not be consistent with known facts about particular individuals.

Collecting hairs isn't pleasant for subject or examiner, especially in rape cases. The ideal hair selection is a good representation of all types of body and head hair. As these are collected by plucking—and as it may take up to a hundred hairs to get full representation from a head, and thirty to sixty hairs from the pubic region for equally representative sampling—the process is quite uncomfortable. Additionally, for victims of crimes in which hair may have been transferred, like rape or other assaults, the clothes should be vacuumed for hairs and head and bodies combed for loose strands.

Trace evidence as large as a cigarette butt is easily seen with the naked eye, but finding trace evidence like hair or fiber evidence may require careful lighting of a variety of surfaces with several light sources. Investigators can't forget to check the *bottom* of nearby objects either. Lightweight fibers float, so they stick to the bottom of items just as well as they do to the top. They may also require time to settle, so a return trip to the scene, especially an enclosed scene like a car trunk, might turn up additional samples.

If the scene itself doesn't yield hair, examiners might be sent to secondary scenes to collect known samples from combs or hairbrushes. Examiners asked relatives of victims of the September 11, 2001, attack in New York City to bring in these items in the hopes of making later identifications. In that case, however, DNA sampling, and not direct hair comparison, was the goal.

MARCEL DUPRIS

"I was reading one of Patricia Cornwell's books the other day—sorry I don't remember the title—but her character, Kay Scarpetta, was busily tracking some crazed guy with a rare disorder that produced baby-fine hair that fell out a lot. I was kinda thinking I wished something that distinctive would jump out of this case, but I rather doubt it."

Dupris slips a sample of hair recovered from a rape victim's body under his scope and stares at it for a long time. "I've never met the victim, but from the feeling the investigators give off, I take it she's a rather remarkable woman. Fortunately—at least I think it's fortunately—I don't have to see her face when I look at the evidence, but that doesn't mean some cases don't stick in your mind for a long time."

Jotting notes to himself, he slips the sample back into its evidence bag. "Serology might be able to get a DNA match. The root looks good." A colleague collects it and wanders out. "Hope so, anyway."

"The oddest cases to us often don't seem that unusual on the sur-face, at least not in the beginning. We had what seemed a perfectly routine case go through here a couple of months ago. An elderly woman was discovered dead in her apartment by one of her neigh-bors. The only family was over in America. The body had started to decompose pretty badly, but there didn't appear to be anything out of the ordinary. The P.M. turned up heart attack—not unexpected in a seventy-two-year-old woman with a known history of heart disease. All was in order. The woman was shipped off to the mortuary where they, in conjunction with the daughter who was then leaving to come bury her mum, decided on a closed coffin service. End of story."

Evidently not, or the story wouldn't still be at the top of the oddest cases list some seven years later.

"The daughter and her family moved back to England a couple years after her mother's death. She was in trade of some type, and it had only been a temporary posting stateside. Shortly thereafter, she meets up on the street with a school chum who says how nice it was to see her mum in France." He pauses. "I don't know all the details, but apparently the investigation started up again when the daugh-ter's husband called the police asking if there wasn't something that could be done about what he felt was a cruel joke being carried out against his wife. Because she'd never actually seen her mother's body—and, I think, because the human mind can be remarkably adept at finding ways for us to avoid pain—the daughter had begun wonder-ing if it really was her mother she'd buried or not. Naturally, she was pretty disturbed, especially when her old friend continued to insist her mother wasn't dead at all. I think the husband was just looking for some way to shut the friend up. But in any case, when the school chum vowed she really had seen the woman, well, the police here got pretty interested."

Even two years later, shouldn't an exhumation be able to deter-mine if the buried woman really was her mother?

"Sure, if we had a body. But the woman, as per her own instruc-tions, had been cremated and the ashes sprinkled at sea. Her husband had been a seaman and he'd been buried at sea."

Meaning there was no evidence one way or the other.

"Not exactly. It's obviously not everyone's cup of tea, but some people have been known to have a pinch of the cremains, the ashes, enclosed in a locket or other item. The husband wondered if there wasn't some way to test them and see if it was her mother." Marcel shakes his head. "Modern science is getting more advanced every day, but we aren't *that* good. The French police were notified of the situation and some pictures were circulated, but I think everyone sort of assumed the issue was dead–until the funeral director showed up with one of the mortuary assistants. Apparently, without the knowledge of her employers, this young woman had altered the usual arrangements for preparing these lockets. There's probably no hard and fast rules about it anyway, but with the daughter not getting an opportunity to see her mother because of the closed coffin, she'd thought maybe she might want a lock of hair and clipped one before the casket was closed. When she heard that it was to be a cremation, she simply slipped the hair into the locket with the ashes before she sealed it."

A fortuitous set of circumstances, but was it useful? Clipped hair wouldn't have any roots for DNA typing, and after it had been sealed up for years in a bunch of ash, would there be any useful information to be gained from the hair?

"All good questions. And, most importantly, even if we could garner something from the hair in the locket, what were we supposed to match it to? This hadn't been a suspicious death at the time, so there were no biological exhibits made."

But the story must have a resolution?

"Oh, it does. Which is how this office became involved again. The old school chum, still being accused of being no more than a troublemaker by the husband, had been plaguing the local French police until they'd finally told her they had more pressing concerns than looking for dead people. She took matters in her own hands, rather literally, by going back to the town where she claimed to have seen the woman and, basically, lying in wait for her, and calling the police when she found her. Though the woman claimed not to be the mother, both police agencies needed more than her word for it. They just didn't know how to go about proving it one way or the other. In the end, she did admit to being the mother. It was right after we'd asked for a hair sample to compare to the one in the locket. We didn't know if there was any point to it, but apparently she thought there might be, and, at that point, also admitted that the body in her apartment was a friend who had no family of her own."

Which leaves the obvious questions of whether or not there was a crime committed and why the mother skipped out in the first place.

"Well, I suppose there was a crime, in that there was life insurance paid out on the assumption that it really was the mother who had died, but I don't think that was her intention. It was almost incidental. The friend had a retirement place, but obviously wouldn't be able to use it; the mother wanted absolute freedom for her remaining years, and I think it was a true crime of opportunity, with no thought to the insurance or anything else. When her friend died in her apartment, it just unfolded in that moment.

"Not one of the most horrific crimes we've been involved in, but certainly one of the oddest. Right up until the moment when I came to collect those hair samples from her, she staunchly maintained that she must be a double. The funny thing is, even today, when we can do more than we could have years ago, I **still** don't know what, if anything, we were supposed to do with either sample. If the dead woman was her neighbor, there was still nothing to match."

The detectives are in the hallway again, and Marcel looks thoughtful. A long Gallic shrug. "But what did it cost us to collect a few hairs?"

Criminalists and forensic scientists may not always deal face to face with the victims of crime, but watching the investigators haunting the hallway and this man peering intently at one of many exhibits—a tiny mass of fibers that *might* mean something to someone down the line—it's difficult to see any difference in the intensity of their interest in the outcome.

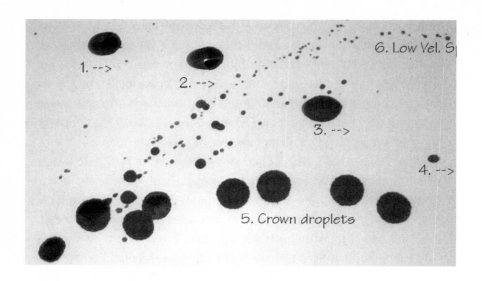

1. -->

2. -->

6. Low Vel. S

3. -->

4. -->

5. Crown droplets

Blood Splatter Patterns

In recent time, blood evidence has become synonymous with DNA analysis. That technology, as well as the more familiar ABO typing, certainly has its place in modern criminalistics, but it wasn't what originally made blood evidence so important. The physical presence of blood droplets—their shape and number, as well as the patterns made when they fell or were sprayed against walls, floor, or ceiling—tells a story.

Photos of splatter patterns are important. The blood itself may need to be collected for other kinds of examination, but wide and close photographic views of the scene and individual blood splatters are taken first. Overall views of all surfaces, including the ceiling, provide scale and relative position. Close-ups of individual drops combine to reveal the overall pattern and the connections between drops—critical information if several patterns overlap one another. The shape and size of drops should be recorded with micro scales clearly visible in the photographs.

The first item the pattern should reveal is the velocity of the droplets. The smaller the drops, the faster they were moving. As blood drops are usually put in motion by a blow or a weapon, with blood pushed ahead of the item making the impact, velocity directly suggests a probable cause of the splatter. A fog of tiny droplets indicates high velocity—almost certainly a gunshot or an explosive device. Large, slow drops often relate to low-impact injuries like a punch. The mid-speed droplets almost always relate to knife or blunt object impacts, though slightly slower, slightly bigger droplets result from a severe beating by hand.

Once something has been determined about the likely scenario, inspection of individual drops can reveal even more. When a drop of blood hits a smooth surface from directly overhead, it tends to form a round spot. Drops from a greater height can form a "crown"—a round drop that hits, bounces up, then falls again to cause small droplets around the original drop. If a blood drop hits at an angle, the edge that hits first—the leading edge—will be round because the droplet was still round at that point. The far side of the now-deformed droplet, however, will be irregular. Looking at the drop, the investigator now knows the direction in which the drop of blood was traveling—toward the jagged edge of the spot.

Next, to discover the angle of a droplet's fall, a little mathematics (the only math you'll find in this entire book) is in order: *The angle of impact is equal to the arc sin of the drop's width divided by its length. Or, A = arc sin (width/length).*

What that means in practice is that, by following the angle of impact back to the origin, the examiner can begin to determine the positions of both the victim and the assailant. General inspection will reveal a simple fact: the longer the droplet, the lower the angle of impact. As all the droplets spread out from one spot, if the examiner can figure the angle for each drop—okay, not all of them, but a representative sample—it's possible to determine where all those angles come together. That will be the exact location where the victim stood, lay, sat, or knelt at the moment of impact.

To determine how close the assailant was, especially if the assailant was within the range of the blood splatter, it's necessary to understand how wounds actually bleed out. A bullet wound can be one of two types: a "through and through" (a wound with two sides—an entry wound and an exit wound), or an enclosed wound (where the bullet enters but doesn't exit, staying lodged in the body). A spray of blood results from most gunshot wounds, regardless of how many holes they produce. The spray on entry is called "forward splatter;" the spray from the exit wound is "back splatter." If there's someone or something between the wound and the surface that would normally catch blood splatter, then blood lands on the person or object, and not on the surface behind them. That break in the pattern fixes the position of a person or object now missing from the scene.

The same methods work with wounds from other weapons, most of which bring the assailant much closer to the victim than a handgun or a shotgun.

Of course, theory is always a lot simpler than practice. In the field, there may be several wounds, all with their own patterns, all overlapping one another and leading to several points. And blood splatter on floors can be destroyed by the assailant, the victim, emergency responders, or anyone who walks through it.

More importantly, there are more types of blood splatter patterns than just impact splatters, and they can all be present at the same scene. Traveling drip patterns result from blood that falls from a wounded person or from an uninjured person who has blood on their clothes, on a weapon, or on their person. Droplets flicked from the blood-soaked hair of a running person leave bizarre splatter patterns. Weapons carried through a scene, especially blunt force weapons and knives, can also leave their own drip patterns.

Further complicating the scene are cast-off patterns—the

blood flicked off a knife, fist, pipe, or other weapon while it's in motion. As a knife leaves the wound, for example, blood adheres to the blade. As the blade moves through space, blood can fly off and land behind the assailant or on the ceiling.

Because human bodies aren't consistent internally—for example, some spots feature arteries full of blood, and others don't—some injuries will display an arterial spurting pattern. Arterial spurting patterns reflect the beating of a heart, forming a wave pattern of droplets that show alternating angle and velocity.

Let's get back to the relatively simple scenario of a single wound with no interruptions in the pattern, and the examiner who has to translate math and angles and velocities into something other investigators, attorneys, judges, and juries can visualize. Until recently, the usual method of demonstrating all this information in a visual, three-dimensional way was to run strings from the blood drops—at the proper angles—until there were enough of them to show where they intersected and where there might be obstructions. That's a long and tedious process.

Small, portable lasers, however, have made the process a little easier. The lights take the place of the strings, eliminating the need to crawl all over the place taping strings to ceilings and walls. Incidentally, it also looks really cool. The very "high tech" appearance, however, tends to make it a questionable technique in jury trails. Though most people are impressed by the light show, many treat it like any other evidence that strikes them as too tech-y and unconsciously tune it out. The stringing process, rarely shown in crime dramas or novels, was brilliantly illustrated in the "Born to Run" episode of *Crossing Jordan*—even though the writers didn't explain how the angle of the strings, which revealed the victim must have been kneeling, was determined. The laser method has shown up several times in the past television season alone. (*C.S.I.* used lasers to establish angles of fire in "Sounds of Silence," though it had nothing to do with splatter patterns.) Patricia Cornwell describes the need for stringing in one of her Kay Scarpetta novels—though perhaps if that story had been intended for film, she'd have updated to lasers, too.

Firearms and Firearm Evidence

Firearms, available in seemingly infinite variety, fill the pages of uncountable books, represented in a detail not even attempted here. Any number of characteristics separate weapons one from the other. From a forensic point of view, both the commonalities and differences provide information. Strictly speaking, the term "firearms" includes cannons, but the majority of criminalists aren't investigating military zones. Therefore, firearms are generally broken into two classifications.

As the name suggests, handguns (also known as sidearms) are designed to be held in the hand. Shoulder weapons, typically called rifles, are designed to be held in two hands and fired from the shoulder. However, modern weapons blur these lines somewhat. Some handguns are so unwieldy that it's hard to see whose hand they were ever intended for, and some rifles are so small that it's easier to use them one-handed.

What all firearms have in common is that something is put in a tube (which is called a barrel), then forcefully projected and intended to hit some target. Sounds simple.

The something in the tube is the ammunition, which also falls into two types: a bunch of tiny objects, or one single, solid object.

The smaller objects—round balls of lead or steel called "shot"—come from a cartridge. The cartridge, usually plastic, contains shot, propellant, and primer. Primer can be found in a ring around the base of the cartridge—rimfire ammunition—or in the center of the base—centerfire ammunition. Keeping all these materials from intermixing are wads, which are pads of fiber—paper, cardboard, or cotton. The primer is lit mechanically

when it is struck by some part of the weapon, usually called a "firing pin." The primer, whether rimfire or centerfire, ignites, burns fast and hot, and sets off the propellant (usually gunpowder) which, in turn, pushes the dozens of small balls out the open end of the tube. Not surprisingly, weapons that fire shot are called "shotguns."

Instead of the dozens of balls of shot, the other type of ammunition has a single, solid projectile in the cartridge. The cartridge in this case is usually metal (sometimes called a "jacket"), surrounding or partially surrounding a metal core (often called a "slug"). Again, there's propellant immediately behind the projectile, and primer directly behind that. And again, a firing pin strikes the cartridge and ignites the primer, which burns the propellant. This forces the projectile—the slug—out the open end of the barrel.

The process of generating enough energy to throw a slug or shot over any distance usually has some side effects. Noise and heat—the waste energy of the whole process—are inevitable byproducts. In some cases, noise can be somewhat controlled by devices colloquially called "silencers." Heat, however, is another issue. The heat of burning primer and propellant acts directly on the slugs and shot, deforming and softening their metals. Shot, being smaller and designed to scatter, is less affected. But slugs, whether from handguns or shoulder weapons, will always be somewhat deformed by heat during the firing process.

Some time ago, a bright machinist realized that a "rifled" tube—one with spiral marks inside it—results in a more accurate weapon. The spirals on the bore (the inside of the barrel) cause the slug to spin, keeping it traveling on a straight course as it leaves the tube. Shoulder weapons with this rifling on the bores were, naturally, called "rifles." Shotguns don't have rifling, but rifles and handguns do.

The heated slug, spinning along the rifling, picks up the impression of the marks inside the bore. If a slug can be recovered at the scene, it will have markings identical to, but in the reverse of, those inside the weapon. Equally important, every slug fired from that weapon will have markings identical to one another. The technical term for the markings are "lands" and "grooves." Lands are the raised areas and grooves are the depressions.

As with all things forensic, theoretical and practical observations can be worlds apart. In a world where slugs are never shot into anything except buckets of sand or, even better from a forensic viewpoint, into water tanks in the lab, every slug from a specific weapon would be identical. In the real world, slugs hit cars, walls, and, unfortunately, people. While slugs are undoubtedly identical when they leave a weapon, they can be flattened into unrecognizable blobs when they smack up against a fire hydrant or a skull. With a good slug in hand, however, it's possible to learn quite a bit about the weapon from which it came.

Gauge Versus Caliber

In the world of firearms, measurement makes clear distinctions between the shotgun and rifled weapons.

Gauge refers to measurements of shotguns only and is an attempt to relate loose shot to an equivalent solid slug. A 12-gauge shotgun packs a load equivalent to a solid lead slug weighing $\frac{1}{12}$ of a pound. A 10-gauge load equated to $\frac{1}{10}$ of a pound. The higher the number, the lighter the load.

Caliber is a measure of the bore of rifled weapons—both handguns and rifles—in hundredths of an inch. A .45 caliber handgun bore is, supposedly, $\frac{45}{100}$ of an inch in diameter and takes .45 caliber ammunition.

Magnum is not a measurement at all, but a descriptive term for the firepower of the propellant. A .45 caliber cartridge's slug and a .44 Magnum cartridge's slug are quite similar in size, but the propellant is very different. A .45 caliber weapon isn't designed for the controlled explosion of a magnum configuration.

Databases on the patterns of spiraling grooves in a wide variety of guns are kept by the FBI; the Bureau of Alcohol, Tobacco and Firearms (ATF); and numerous other agencies. Consequently, the number, spacing, and direction (clockwise or counterclockwise) of the lands and grooves on a slug—in combination with the caliber of a slug—can tell an examiner exactly what brand of weapon was used.

Recovering slugs involves locating them in objects at the scene. This, of course, includes people—both living individuals (usually in a hospital setting) and deceased individuals (at the morgue). Most medical staff understand the importance of retrieving slugs or their fragments without doing further damage to them (using hands or rubber-tipped forceps for retrieval), and all the investigator needs to do is take official custody of the evidence at the time of examination or autopsy. Recovered slugs should be allowed to dry and, if there are more than one, packaged separately. Retrieving slugs from walls, doors, or tires should only be attempted by examiners aware of the details that need preservation. Whenever possible, embedded slugs (and any object in which they're embedded) should be taken to the lab *after* any on-scene work—such as determining angle of fire—is complete.

Of course, while slugs or shot come from one end of the weapon, physical by-products of what is really a chemical reaction are released from the opposite end. The most obvious by-product, the cartridges, reveal as much about the weapon as do the slugs. As already noted, tool marks can be matched to individual tools, and the firing pin of a firearm is nothing more than a tool. The breech face that rests against the cartridges also creates marks and, if a weapon automatically ejects cartridges, even more comparable marks are added to the cartridge. Close inspection and detailed photographs or micrographs of the marks left on the cartridges allow later matching with suspect weapons. The pattern of fallen cartridges can reveal where shooters stood, so detailed documentation of the location of every spent cartridge should be attempted before cartridges are collected and packaged, one cartridge to a container.

WANTED—FIREARMS/TOOLMARK EXAMINER

The Las Vegas Metropolitan Police Department is seeking applicants for the position of Firearms/Toolmark Examiner.

Qualifications include a bachelor's degree in criminalistics, forensic science, chemistry, biology, or related field, and three years of responsible research and practical experience working in a forensic laboratory as a professional firearms/toolmark examiner.

Responsibilities include performing scientific and laboratory analyses on firearms and tool mark evidence, interpreting test results and forming conclusions, preparing reports, and testifying in court as an expert witness.

Salary: $52,035-$77,029/annual.

DRUGFIRE

Drugfire is a computer database program that stores images of fired cartridge casings and bullets—either from the scene or from lab test firings—and recovers possible matches for comparison by examiners. When an apparent match is found, the actual evidence is procured for further analysis. Hundreds of images are added to this database daily, and the number of matches made in jurisdictions like the Minneapolis Department of Public Safety doubled from 1998 to 1999.

Because Drugfire's data crosses all geographic borders and cuts across all jurisdictions, investigators are discovering the interstate matches (once difficult to track) that can establish a criminal's repeat behavior patterns for sentencing boards.

Not all firearms eject spent cartridges. But, without exception, all release GSR—Gun Shot Residue—the cloud of vapors and particulates resulting from that controlled explosion. GSR contains specific amounts of several chemicals, including lead, barium, and antimony, the material of primer mixtures. Finding it on suspects means one of three things: they fired a weapon, they were present when a weapon was fired, or they handled a recently fired weapon. The same elemental metal analysis that sorts out the content of GSR can also determine the mix of metals in slugs or shot. Many manufacturers deliberately include metal additives in known proportions, so tracking individual batches of ammunition becomes a possibility. Cartridge casings vary less, but every scrap of information can help exclude areas of search, allowing investigators to concentrate on more promising leads.

With so much potential, investigators actively search out firearm evidence, carefully collecting items as large as the weapons themselves and as tiny as the micro-trace of metals on a suspect's hands. Though not explosives in the usual sense, firearms and firearm ammunition pose real hazards to investigators and lab personnel.

The first hazard is obvious: any loaded weapon is a danger. And any weapon should be presumed to be loaded until proven otherwise. "It's hard to believe, but we once had a firearm go off on a countertop in the lab," recalls Ballistics Examiner Françoise Gillaume. "No one touched the box, it just went off. Two things at work there. One, the thing should never have been transported to the lab loaded. Two, it should have been clearly labeled as a weapon. We reviewed some policies very shortly thereafter."

Not all firearms can be immediately rendered harmless. Older weapons, especially those recovered from saltwater, often suffer rust and corrosion that make it impossible to open and empty them, but not necessarily impossible for the firing pin to fall. In such a situation, special precautions should be taken. Specifically, such weapons should be kept in containers that have impermeable walls, and an appropriate means of securing the weapon and preventing motion should be devised.

Weapons aren't the only danger. "Old ammunition itself is a real problem. Totally unconnected with any crime—except perhaps stupidity—we were called out to a shed riddled with bullet holes. A box of .38 cartridges was left in it, in among a bunch of chemicals—mostly unlabeled, of course. Something leaked, corroded the cartridges, one went off, then the next, and then the whole box!" Because hand packing occurs fairly often, especially with shotgun shells, it's very possible that investigators can encounter loose supplies of hazardous components.

The Don'ts of Firearm Collection

- Don't transport loaded weapons, for obvious reasons.
- Don't pick up weapons by sticking a pencil or chopstick down a barrel. It can alter the microscopic characteristics of the lands and grooves, ruining it as evidence that might be matched to particular slugs. A weapon can be lifted by inserting an implement like a pencil behind the trigger guard.

- Don't print any weapon without consulting the firearms examiners—and especially don't print by cyano-fuming. Cyano fumes can coat the interior of a weapon, as well as the outside, and deform evidence. Sealing the bore, chambers and any other interior opening before fuming allows everyone to do their job, while preventing destruction of any evidence.
- Don't handle cartridges. They can contain prints evidence.
- Don't assume any ammunition or ammunition components are chemically stable.
- Don't use paraffin tests to check suspect hands for residue. It's no longer considered credible and has been replaced by the much more accurate metal assay. Gunpowder includes nitroglycerine and other organics, which means MECE (micellar electrokinetic capillary electrophoresis) can be used to match gunpowder—and its residue—to particular weapons. Few jurisdictions have an SEM (scanning electron microscope) just sitting about, but such a scope is also a dandy tool for analyzing GSR.
- Don't package metal slugs, weapons, cartridges, or shot in plastic containers. Any dampness present may start corrosion.

Questioned Documents

Though fraud is undoubtedly a crime, questioned documents—such as autographed baseballs, challenged wills, and forged checks—seldom arise during crime scene searches, can usually be pursued with some leisure, and are usually only examined in the lab.

Ransom notes at kidnappings and demand notes from robberies, however, require swift analyses on many levels.

THE FULL SCOPE OF
QUESTIONED DOCUMENT EXAMINATION

Though the project we're looking at here deals primarily with one aspect of document examination, the field is considerably broader and includes several activities:

- distinguishing the forged document from the genuine

- detecting erasures or substitutions on documents

- restoring erased or obliterated writing

- analyzing inks, papers, and chemicals involved in handwriting or document creation

- attributing handwriting, signatures, printing, and other writing to individuals

"In ransom or demand notes, it's less a forgery issue than an attempt to disguise handwriting samples," says Frank Perry, who has been examining documents for nearly thirty years. "A forgery is an attempt to match someone else's handwriting. Disguised handwriting tries to *not* match a particular person's writing, which is, of course, why it fails."

Fifty years ago, when Frank was a child, some very high-profile kidnappings were taking place. In the summer of 1956, Frank was eleven and Peter Weinberger was just a month old. Frank was listening to his neighborhood buzzing with the news that a child had been kidnapped. Peter Weinberger was dying. "I can't say it affected my career choices, but I guess it always stuck in my mind that they identified that child's kidnapper by handwriting comparison. It was incredible to me that anyone would even *start* working through literally millions of public records and samples for some obscure writing aberration. For the adults, there were echoes of the Lindbergh case, and I remember the deep-seated dread of that time."

> He held out the sheet of paper that he'd
> found in his pocket when he'd been looking
> for baby-sitter money. "I noticed there was
> tremble in his handwriting. That's what
> happens when somebody tries to disguise
> their writing. I remembered it was Hardy
> who'd written down what I dictated but why
> would he try to fake his writing? There
> was only one reason—because he'd written
> the extortion note. I checked the
> lowercase i in 'two miles' and the dot was
> a devil's teardrop. That confirmed it."
> JEFFREY DEAVER, *The Devil's Teardrop*

The Weinberger case is an example often used in response to those skeptical about the value of document examination. "Handwriting examination isn't 'handwriting analysis' or 'graphology,'" Perry explains. "When I sit down with a document, I'm not trying to equate a big, swirly style with an egotistic temperament. I'm not interested in that sort of question. I'm interested in, first, is this real writing or disguised writing? Then I want to know if there are distinctive writing habits, like in this sample you provided, of making q's for both q's and g's, something that can be found in other samples and linked back to this exemplar. I don't know if making q's instead of g's tells me anything about your personality. Maybe you missed school the day they did 'g', I don't know. And I really don't need to know that."

Margaret Rumsfeld, a forensic linguist, is also quick to separate linguistics from graphology. "Linguistics is the study of language and its use. Regionalism means that English is different in New York than in Louisiana. Individually, people have quirks in their use of language. In the Unabomber case, linguists read the bomber's manifesto for patterns in the use of words, favorite phrases, even consistent spelling errors. The bomber's brother, on reading the published manifesto, contacted authorities because of similarities in both content and language style to his brother's previous writings. Most people writing as themselves have a distinct 'voice.' Psychologists can make determinations about state-of-mind from what individuals say, whether orally or in writing, but I don't. I certainly wouldn't think it's possible to categorize an individual's whole personality based on whether they put little hearts over their i's.

"What I can do—given two pieces of writing, two letters, two articles, or two manifestos—is note any similarities in style, use of language, or phrasing. From that, I can sometimes determine the likelihood of those two pieces being written by the same person or not."

Frank Perry seldom has documents as long as the Unabomber's manifesto to work with. "Demand notes can be a short as 'Put the money in the bag!' And, really, there's only so many variations on it." Holding up copies of ten bank/store robbery notes, he makes a quick word count. "Less than a hundred and twenty words in total, not nearly enough for Margaret's purposes."

TEXT OF THE ORIGINAL LINDBERGH RANSOM NOTE

Dear sir:

   Have 50,000 $ redy 2500$ in 20$ bills 15000$ in 10$ bills and 10000$ in 5$ bills. After 2-4 days we will inform you were to deliver the mony. We warn you for making anyding public or for notify the polise the child is in gute care. Indication for all letters are singature and 3 holes.

All spelling errors, symbol placements, and punctuation are original to the document. What is not visible in a text transcription, what examiners like Frank Perry need, are the physical papers and inks that formed the note, the images drawn on it (in this case some believed a series of dots had symbolic references to targeting systems, while others thought them encoded initials referring to Bruno Haupt-mann), and the chemical or impression trace they might carry.

Linguists like Margaret Rumsfeld might note the placement of dollar signs after the amount (a European tendency instead of the North American standard), and the basic lack of punctuation, and suggest the possibility that the writer's first language was something other than English, but even for that, she'd likely defer to Frank because, "the most common linguistic fraud is to pretend an unfamiliarity with English. The stroking in handwriting, whether the individual stopped to consider whether or not to leave out punctuation—that type of information comes from the document examiners."

DAVID GOTTY

Carla is left-handed—the only left-handed person among the support-staff personnel in her department—and she is, once again, writing reams of samples for David Gotty. "I tend to treat it as therapy," Carla laughs. "I write all this *stuff*, everything that ticked me off that day, knowing full well that Dave doesn't really *read* any of it. I wrote 'Will you marry me?' nine times yesterday and he didn't even notice!"

"Not true. I noticed." Dave seldom seems to straighten up completely—all the better to peer through the dual scopes set up on his short piece of counter space. "I also noticed the odd lifts when you wrote it, completely dissimilar to everything else on the page and properly deduced that you weren't nearly as serious about marrying me as you were about shooting your insurance agent or re-membering to pick up your dry cleaning."

"You are one scary dude, Dave!"

Perhaps, if you happen to be a criminal, that would be true, but David Gotty describes himself as "a quiet man of letters—all twenty-six of them."

"That's basically what I do, compare one uppercase A with an-other, then the B's, then the C's, and so on. And then the lowercase letters. Numbers, too, sometimes, but mostly letters. In general, by

the time I've got those covered, I've got a pretty good handle on whether I'm looking at work by the same individual or not."

And the lifts?

"Lifts are those points where the writer lifts the writing implements from the writing surface. I haven't yet quantified it, but in some people, not just left-handers like Carla, there's a tendency to have more lifts in writing that reflects some sort of untruth. Carla, for example, is happily married to a much nicer man than me, so I rather doubt she was writing seriously when she asked me to marry her! There were a *lot* of lifts in there!"

Stretching his arms above his head, David Gotty transforms from a question mark into an exclamation point. Clearly, I've been looking at too much handwriting. But I'm sure David would appreciate the comparison.

"My theory is that when a person knows exactly what they want to say, they do it in a single long thought. They don't pause and think about what they're going to say next, they just write. When they're inventing, or playing at being someone else, or deliberately dissembling, they stop to think, and they stop writing." Passionate about his work, Gotty blushes when forwarding theories he has yet to "quantify" to his own satisfaction.

His attention to lifts paid off in a recent case, however, and pointed police in a completely different direction—the right direction.

"Often, with questioned documents, the police can't get exemplars of suspect writing that we can use. Either the suspect isn't in custody, or simply refuses to give us a sample. And, not all samples are equally useful. So if we have a questioned sample of block printing, and no block printing in any of the suspect's exemplars, we're pretty much out of luck.

"That was the situation when Arno Quilty and Tom Pynn were charged with grand theft in a bank robbery. Quilty later died in custody, and Pynn's attorney argued hotly that the confession Quilty was said to have signed before he died was a forgery and could not be submitted in a trial against Pynn.

"We had exactly one sample of Quilty's writing, the demand note in the robbery, which—surprise!—wasn't signed."

No points for comparison.

"Right. But, there's more to this game than just matching letters."

Lifts?

"Among other things, yes. Handwriting varies with age, with illness, and, most notably, under stress. Human beings are pretty consistent in how they respond to stress."

More lifts?

"Usually. And, equally important, there's line quality. The smoothness of the path of the implement across the surface. Jaggedness, tremor, and angle of contact all increase when we're nervous, frightened, excited, stressed. With that in mind, we looked at the two samples we had and tried to determine when Quilty would be more stressed—in the relative calm before the robbery, when he had the opportunity to write a dozen practice notes if he was so inclined before going to the bank, or later, in police custody, when he was deep enough in undisclosed insulin shock—he died of diabetic complications just hours after he supposedly signed his confession—and facing trial and conviction?"

Neither sounds particularly restful.

"No, but, there are degrees of stress, and we had witness accounts of Quilty's behavior throughout his capture. Until he actually went unconscious in his cell and later died, he was highly agitated, almost manic, desperate to talk to Pynn."

Obviously more stressed than he would have been prior to a job he knew he could walk away from at any time.

"We think so, and so did the forensic psychologists we consulted with before looking at the samples." He grins a little at the memory. "So, when we looked at the two samples, and found a firm hand, good line quality, and almost no lifts in the confession signature, I had no second thoughts about whether there was doubt as to the identity of the writer. Unless Quilty had nerves of steel, which had been noticed at no other point in his life, he did not sign that confession."

Length isn't the only difference between material sent to Margaret and material sent to Frank. "Everyone prefers to work with original documents," says Frank. "But while Margaret can work with copies, document examiners must have the real thing if we're to make the most of our opportunities. Quite often, photocopies come to us, or get faxed to us, but they're practically useless. We get less than 10 percent of the information from copies than from the real thing. The general public— even many law enforcement people—doesn't realize documents are three-dimensional and we need every dimension we can look at to solve immediate crimes."

Not surprisingly then, the first thing many examiners investigate is the paper itself. On the gross scale, impressions like watermarks help identify brand and type of paper, which can then be traced through retailers. On the micro scale, paper, a malleable material, picks up marks from the surfaces it rests against or that rest against it. Frank recalls a 1982 case when he found an anchor image pressed into one corner of a page. "I thought first it was a watermark, but it didn't look right. Really, it was a mark from a jacket button, a sleeve button we eventually tracked to sales by a uniform manufacturer."

Fabric impressions rarely occur, but are worth searching for. "The imprint of the tape in a lawn chair seat appeared under UV investigation of a notebook recovered from the home of a missing woman we were investigating. There were no lawn chairs at her residence, but there were at her boyfriend's. She'd actually been assaulted at his place, but he'd brought her things back to her apartment to hide the fact she'd ever left for his place."

Other important impressions are the writing markings left on underlying sheets. In very few cases are the notes that are received by victims the first effort of the ransomer or thief; even short notes get rewritten dozens of times, either in an effort to further disguise the handwriting or to change word choice. "So, a note may have the impression of earlier attempts, or sheets below the one received can have evidence of that note."

Like other impression marks, side lighting often reveals unseen imprints. (No one rubs it with a pencil like they do in the movies!) More unusual treatments can pull up even further evidence. Watermarks appear when viewed with transmitted light (light shining through the paper). The light may be ordinary white light, filtered light, or light in invisible wave lengths, like infrared or ultraviolet. Each frequency of light may reveal more impressions. ESDA (electrostatic detection apparatus), which works much like the electrostatic devices used in lifting prints, provides examiners with a record of impressions that are easier to analyze than paper itself. Photographers using oblique lighting and high-contrast films (occasionally IR or UV film can be used) develop useful images, as well.

UP IN SMOKE?

On top of being a favorite way to dispose of ex-lovers, incineration is a natural method for destroying paper evidence. More than one criminal has attempted to eliminate the document trail by burning away the evidence.

Frank Perry spends many work hours trying to preserve documents that have been reduced to curls of ash, but says that a snippet from Jeffrey Deaver's *The Devil's Teardrop* is as sound an explanation of physical preservation and reconstruction as most descriptions he's seen in reference texts. "There are chemical aids but, in this particular novel, the fictional examiner's field kit was burned, so he couldn't use parylene or any of the other stabilizers that might have helped. Instead, the author had the character fall back on older methods—getting the ash, remnants, and paper shards between glass and back to the lab ASAP."

Parker finished cleaning the glass and turned his attention to the evidence . . . to his dismay . . . much of the ash had disintegrated. . . . Still it would be possible to read some of the unsub's writings on the larger pieces of ash. This is done by shining infrared light on the surface of the ash. Burnt ink or pencil marks reflect a different wavelength from that of the burnt paper and you usually can make out much of the writing.

Like tool marks, machine impressions—from typewriter keys, roller parts, printer heads, and even photocopier transport parts—can be matched back to mechanical devices that were in contact with the paper.

Though some modern printer cartridges combine inkwells and print heads in one disposable unit, other equipment and supplies provide opportunities to match evidence. Typewriter ribbons, plastic film cartridges, correction tapes, fax film, adding machine tapes, and carbon sheets are found in many home offices, and all can be examined for impression and tool mark evidence. One of the oddest impression mark possibilities may be the transfer of paper impressions to typewriter and printer ribbons. Essentially a specialized fabric, paper can transfer an imprint of itself to the ribbon in the same way marks of the ribbon (another fabric) transfer to paper.

The voluntary submissions made by paper and ink manufacturers add to the paper and ink databases maintained by the FBI and dozens of regional jurisdictions, so examiners stand a good chance of matching one or both of those elements. Paper characteristics include size, thickness, rag content, reflectance, color, and opacity. Inks vary in composition, and their recipes often change; but like paper, the formulations have distinctive colors and physical characteristics.

Knowing which paper and ink were used to produce a document narrows the possible location or origin—maybe even to a very small region—but doesn't prove who did or didn't write on it. For that, handwriting or hand printing examination is required.

"The two scenarios for handwriting examiners are one—trying to match an unknown to a known, Q to K—or two—to compare two unknowns to one another, Q1 to Q2," says Frank. "Unknown to known might be a situation like the Weinberger case. Investigators had a sample from an unknown subject—the ransom note—and they were trying to match it to samples from known individuals in public records. Autograph examiners work on this principle, comparing a suspect signature to verified, well-documented autographs. The scenario of unknown to unknown puts the examiner in the position of determining if two documents, both source unknown, came from the same unknown subject. Later, of course, they might well be asked to attribute both documents from the unknown to a third example from a known subject and decide if all three were created by one hand."

To begin work on a document, examiners go back to Frank's first question, "Is this a real exemplar or is it an attempt to disguise real tendencies?"

No two examples from the same person will be identical. Within the range of one person's writing, however, there will be consistencies. Brains are basically hard-wired for repetitive tasks. One person may cross individual t's in the word "tort," while others wait to finish the word and draw one long line through both t's. People develop habits, even in handwriting.

"Think of how odd it feels to put on your left sock first if you always put on the right one first. Or how uncomfortable you feel if you forget to put on your wristwatch, or don't wear your wedding band. Smoking isn't the only habit tough to break. Actions can be conscious or unconscious, deliberate or automatic. When we examine an example of handwriting, we look for internal consistency. A high degree of internal consistency suggests this is the person's normal writing habit. A high degree of internal inconsistency suggests attempts at disguise."

To demonstrate that consistency, he hands me a typing sample collected at the same time as the writing samples. "See? You type as much, likely more than, you write in longhand, but the same traits are evident. You type everything with q's, too. Must drive your spellchecker nuts."

He's right. It does.

"Your writing is consistent internally, even consistent with your typewriting habits. Given these samples, I'd believe the exemplars you provided are very much in your usual hand, no disguise, no tremble or hesitation, and I'd certainly be able to identify other examples of it. Not just on the basis of that 'q', either!"

Disguised writing exhibits numerous contradictions. "The most common attempt to disguise handwriting is to use the opposite hand. It switches the slant of the writing, but over any stretch of writing, you'll see the slant begin to shift back to the writer's natural stance, regardless. Writing takes manual strength and dexterity. It's tiring to force muscles into unnatural positions and, unconsciously, the writer searches for a way to ease the discomfort." The same discomfort will interrupt the flow of writing, creating many more spots than usual where the pen or pencil is lifted from the page.

Right-handed writers who switch to their left also hold the pen at an unnaturally high angle to the page in an attempt not to smear their work. "True lefties don't need to exaggerate the arch of their wrist. They know how to keep their cuffs clean and maintain a single angle throughout rather than constantly overadjusting." Oddly enough, left-handed individuals don't use this particular subterfuge as often as their right-handed comrades. The truly ambidextrous person—a rarity—doesn't use it, either.

"Feminization is a tactic popular among males. They'll round out their writing, add little flourishes and fillips to it, add strange things to the dots over their i's. Women may do the reverse, produce more angular writing, more pressure on the page, bigger overall writing." Frank hasn't done scientific studies to prove it, but he suspects such individuals are unconsciously attempting to recreate the handwriting pattern of a parent. "We mirror our role models in other areas, so it's a pet theory of mine. Maybe I'll get time to actually work up a research protocol some day."

He fingers the sample papers on his desk, pulling out a sheet of lined looseleaf. "Most forgery starts on paper just like this. The first thing I ever forged was my mother's signature on a test I'd flunked—I imagine most kids faked notes like that, or excuses for skipping school. My sister wrote them for cigarettes for us when we were kids, and we'd go three or four blocks over to get smokes from a storekeeper who didn't know our folks. We're basically creatures of habit."

What *is* known is that, short of cutting letters out of newspapers, the most common way to disguise handwriting is to resort to hand printing. Few adults print unless requested to—as in the "Please print" instructions on income tax returns. It's just too slow. Preexisting comparison documents of printing samples will, naturally, be harder to locate. And printing in all capital letters thwarts the tendency among adults to get lazy and begin stringing the letters together.

Despite all these efforts, certain characteristics come through, and experienced examiners soon become familiar within a single document with a writer's stroke, inclination, and other identifying characteristics. Given two documents, both are subjected to the same scrutiny. Examiners note any absent characters, or characters that are in one sample but not in the other, then occasions of similar characteristics, then number of dissimilarities. From these comparisons, they can report on the likelihood of the two documents being written by the same individual.

In *The Norwood Builder*, the slightly mad genius Sherlock Holmes presents his trusty sidekick Watson with a handwriting sample and orders him to deduce what the various degrees of penmanship—ranging from excellent to messy, yet all found in one document—might say about the writer, if indeed it was all written by the same individual.

Watson, after a moment's thought, is delighted to announce that the writer, definitely a single writer, must have been "working in a moving vehicle!"

Holmes, of course, one-ups him by declaring, "Of course!" and then proceeds to explain how the writing—excellent at the stations, poor while in motion, awful when rattling over intersections of the tracks—proved the author was or wasn't on an express and, based on that, which train he must have taken to London.

To comments that the job of document examiner is declining in importance, what with printers and laser writers taking over document production, Frank just shakes his head. "Most ransom notes and most demand notes worldwide still get cranked out by hand." Looking back to the summer of the Weinberg kidnapping, he adds, "They *had* the guy back then. They had him dead to rights, from handwriting examination. They just couldn't get him fast enough. I figure this technology, which is the bane of most examiner's lives at the moment, has got to start working for us eventually. Until then, we need more examiners, not fewer."

Experience clearly counts in document examination, but training gets investigators up to speed faster. Training also contributes to more uniform ways of reporting, which makes information sharing faster, too. One of the best training programs is run by the FBI. Candidates for the course have to qualify through the usual FBI employment process, then prove that they have "acute vision, no color blindness, and the ability to lift fifty pounds." Then they can start learning. It's not a quick program. It's two years of classes and then fieldwork, then more classes and more fieldwork. They write numerous reports and opinion defenses and participate in no less than five mock trials before being certified. And those mock courtroom sessions are no cakewalk. Not with some of the best examiners in the world breathing down their students' necks, absolutely ready to throw the toughest questions they can think up.

Training at the FBI does have its compensations, however. First and foremost, trainees are granted access to some of the greatest document collections in the world and, no less importantly, access to the FBI-maintained document examination databases, including the following:

- **Watermark File:** Collections of manufacturers' information and designs used in watermarks and unusual imprints, such as monograms.
- **Anonymous Letter File/Bank Robbery Note File:** Information, examples, and images of numerous letters of threat and demand.
- **Checkwriter File:** Checkwriters (those machines that fill in amounts on checks and money orders, often leaving holes as well as ink) have defining characteristics that are identified in these files, often with accompanying samples.
- **Fraudulent Check File:** A fake check contains dozens of pieces of information, and this collection cross-references examples and incidents by company names, beneficiary names, and signatures on counterfeit checks, traveler's/bank checks, and a variety of money orders.
- **Office Equipment File:** Information on typewriters, typefaces, printers, fax machines, photocopiers, and other office gear, as well as accessories and disposable parts.

Forensic Entomology

It goes without saying that staid science often bows to visual value when Hollywood takes on real world technical trials, but "Sex, Lies and Larvae," a first-season episode of *C.S.I.*, provides a perfectly convincing view into the world of forensic entomology. Not only did fictional investigator Grissom collect bugs (all sorts of bugs at all stages of life), but he also measured temperatures and raised his larvae. (He even had to wait the requisite number of days for adults to appear. It didn't happen overnight!)

If a few bugs hatching on TV seems to cause an unwarranted amount of enthusiasm, it's likely because, of all the investigative techniques brought to the screen, this is the one that gets the least—and worst—play. One truly dreadful film suggested not only that the adult flies were the only ones of use, but also that all required of the forensic entomologist was to identify a

species. The investigator then promptly announced that the corpse—in a *sealed* oil drum coffin—had been there for precisely eight days.

Sadi Khee only wishes it were that easy.

"If all I had to do was wave a bug net, I'd be out of a job because butterfly collectors are so much better at it than me!"

Forensic entomology, the law enforcement side of the study of insects, was traditionally reserved for determining a PMI (postmortem interval)—another way of saying time or date of death. That's no longer true. Insects are contributing to establishing suspect movement in cases where there are no bodies at all, the point of origin in many illegal drug caches, and the path of stolen automobiles as they get handed off in organized car rings. It'll be some time before insects are collected as routinely as fingerprints, but their uses at death scenes have become established well enough for most investigators to have at least a basic knowledge of collection standards and methodologies.

First, however, a brief lesson on the love lives of bugs is in order. Insects are never born in fully developed adult form.

Instead, they're dumped into the world either as eggs or larvae. In either case, each species of insect progresses through the other stages of its life cycle at predictable times. Unlike human children, which lurch from adolescence to adulthood when we least expect it and without much outward sign, insects turn from egg, to larvae, to pupae, to adult with abrupt and predictable changes. Most people are familiar with the caterpillar, which becomes a butterfly; entomologists recognize these stages with quite a bit more detail.

Blowflies, one of the most useful species to forensic entomologists, have well-understood life cycles that progress reliably from egg to first instar (a stage of larval growth, usually determined by maggot size and how often it has shed its skin) larvae, to second instar larvae, to third instar larvae, to prepupae, to pupae within puparium, and on to imago. For example, a larva just crawling from its egg might be 2 to 5mm in length. It will happily munch away until it gets too big for its britches; then it will rupture, leave its old skin behind, and start munching some more. At that stage, it's a second instar and perhaps 6 to 9mm long. After shedding its skin a second time, it's third instar, and maybe 10 to 13mm. In the blowflies, the third skin begins to darken, thicken, and toughen. It's considered prepupae when it is still capable of moving but has given up the hunt for food in favor of the hunt for a good site in which to settle and get on with the business of maturing into an adult. That thick third skin becomes the puparium (similar to the butterfly's cocoon or chrysalis), and inside it the larva undergoes one more partial molt, then becomes a pupa as adult features begin development. When the third skin splits, the critter crawling free is the imago form, the adult. It will not grow any more. Small flies aren't younger than larger flies, they're just flies of a different species altogether, or two flies of the same species but of different sexes.

The length of time a blowfly spends at each stage has been studied in some detail, under a variety of conditions. For example,

most insects develop more quickly in warmer conditions than cool. The forensic entomologist can roughly estimate the length of each stage under a laboratory's controlled conditions, then adjust the schedule to account for actual conditions at the scene. (More on that below.) Before determining how long a body has been dead, as opposed to how long insects have been infesting it, the entomologist must first have a good idea of when the first blowflies arrived and laid eggs on the body. In the case of blowflies, the preferred time is approximately two days after death. So, if the third instar would normally appear on the eighth day after eggs were first laid, and conditions are close to normal, the person would actually have died *ten* days prior to discovery.

Obviously, blowflies aren't native to all areas. Other species are also studied so that experts can discover their cycle. Not all insects are attracted to bodies at the same time. Some prefer the freshly dead, others the well decayed. Some don't actually feed on the bodies at all, but on the molds forming on them, or on other insects that have been attracted to them. New arrivals that appear together, or at the same time, are sometimes collectively referred to as waves, and a forensic entomologist would be able to gauge the normal progression, species by species, for his or her region.

In determining when a wave appeared, it's not sufficient to say "two days after death" or "at a fresh corpse." That assumes a set of normal conditions, and that every body decays at the same rate. Neither is true. Frozen bodies decay at quite different rates from bodies inside oil drums or bodies staked out in the middle of the Mojave Desert. The entomologist's timeline must account for these differing rates of decay. To make that possible, bodily decay is described in quantifiable terms.

But even in the absence of outside mold, bacteria, worms, or insects, bodies rot from the *inside*. Like the life cycle of bugs, this process follows a well-documented path.

STAGES OF BODILY BREAKDOWN

Initial Decay: Externally, the corpse appears much as it did in life, but decomposition has begun due to the actions of "bacteria, protozoans, and nematodes" *already present* in the body when it was alive.

Putrefaction: Gas formed by the activity of organisms within the body cause it to swell—and smell.

Black Putrefaction: A bit of a misnomer, actually, as the characteristic discoloration of the flesh accompanying this stage may be blue, green, purple, brown, or black. The swelling of the previous stage collapses again as that gas begins to escape. The swelling decreases, but the smell increases dramatically.

Butyric Fermentation: Tissues and organs have become fluid, fluid has escaped by a variety of means, and now the body begins to dessicate (to dry out). Mold usually covers some or all of the exterior. A different odor—not good, but not as "knock you over and send you gagging" as the previous one—is noticeable.

Dry Decay: Not mummification, but a slow process of continuous decay, during which time the tissues continue to rot, dry out, and shrink until skeletization has occurred.

Being able to associate a particular stage of death with a particular type of insect infestation also makes it possible to determine if a body was *unavailable* for insect infestation for some reason. If blowflies like to infest bodies in the putrefaction stage and an examiner finds a body well past that stage, in an area where blowflies are habitually found, and there's no sign of them or their remains on the body, the entomologist will consider the distinct possibility that the body was elsewhere at that time in the corpse's history. Maybe it was moved from a sealed area after putrefaction was well advanced? Maybe it's winter and blowflies aren't around? Maybe the body went through that stage at night when blowflies are inactive?

Obviously, a lot of factors are at work simultaneously around bodies. An entomologist must be intimately familiar with the insects of the region where a body was found or where a death occurred—not necessarily the same place! He or she must know the insects' life cycles, their seasonal activity, and the multitude of other influences affecting bodies and their rates of decomposition.

Obtaining a firm picture of the conditions where a body is found is essential to understanding what may have been affecting the corpse and the insects around it. General scene photography will concentrate on the body, situating it in the scene, but will often miss the details forensic entomologists need. As a result, their examinations will likely begin with their own series of photos. Adult insects of all species in the vicinity, evidence of insect activity in any standing blood (an entomologist can help discern changes in the splatter patterns made by insects and not representative of the original droplet, helping to eliminate odd findings), and the remains of insect activity (feces, egg casings, cast-off pupae and larvae sheddings, and dead adults) must be photographed.

Body parts showing insect infestation need photographing before the body is removed. The days of sluicing maggots and eggs down the drain as a nuisance to the autopsy process are hopefully over, but because these are living specimens, it's entirely possible that the entomologist's evidence could fly away at some point between the discovery of the scene and the autopsy. Additionally, some forms of insect damage have, historically, been mistaken for criminally inflicted bruises and abrasions.

Next, a bunch of thermometers are set in place around the

scene. For eggs and larvae found in masses on the body or in nearby ground, thermometers should be inserted at the center of the mass and temperatures should be recorded. The basic measurements are air temperatures at ground level and about five feet above that (in both shade and sun if the body may have been exposed to both), and ground temperatures at the surface under the body and at the skin surface of the body. If the body is outside in a nonurban setting, temperature readings just below any leaf layers (and at a depth of a foot or so) should also be taken.

While all those thermometers come to equilibrium, the actual collection of insects and insect remains from the body and surrounding areas can begin. A live-specimen container with a known temperature and something for the insects to eat (raw liver, for example) are essential. It's extremely unlikely that only one species of insect will infest a body and distinguishing between the first, second, and third instars of maggots, even in just one species, isn't always easy. The best way to determine species and life stage is to raise the eggs, larvae, and pupae to adult, carefully noting how long it takes for them to progress from stage to stage. From these notes, it will be possible to work backward to the time/date of infestation, add the likely death-to-infestation period, and arrive at a very good time frame for death.

Ideally, examples of all insect remains and preserved samples of all insects, in all stages, should also be collected. As some insects prefer the fluids leaking into the soil instead of the body itself, collecting soil samples from directly beneath the body and from the surrounding ground remains an important part of the process. If very few samples are available, the entomologist must make a decision on the spot as to how live/preserved specimens should be allocated.

With live and preserved samples secured, the documentation of the scene continues with descriptions of impressions gathered. Were insects found where they were expected? Where they weren't expected? A child in 1943 was poisoned with a

preparation taken by mouth. When her body was discovered nine days later, insects had infested the usual areas—nose, eyes, and other body openings. All except her mouth. A curious medical examiner—there was no entomologist either on staff or on consult—couldn't find a poison through toxicology, but after nine days the poison could have broken down in the child's body. The lack of bugs around the mouth continued to bother him enough to mention it to the investigating officer, who "leaned" on the suspect. Luckily, the suspect believed it was possible to make a case with that information and promptly confessed.

As mentioned, a lack of expected insect infestation might indicate a body was moved. And so might the presence of insects not normal to the region where the body was found. Not only are insects predictable in their reproductive and growth habits, they're predictable in their preferences, as well. One species might prefer an urban habitat, others rural. Bodies in water, buried, or in open air might well display different patterns of infestation. Wounds attract attention to areas of the body that are not usually primary infestation sites. Large numbers of maggots—or maggot damage—on the hands (not a typical first infestation site) might suggest the presence of defensive wounds otherwise hidden by decomposition.

Weather descriptions from local observers, meteorological stations, and other sources that cover any periods when the body lay outside round out the on-scene observations, letting the entomologist make any adjustments to the life cycle equation.

THE SINCEREST FORM OF FLATTERY

In the season-two premier, *C.S.I.* continued its preoccupation with six-legged murder witnesses. In "Burked," an investigator scratches his way through the second half of the episode after being bitten by chiggers at the scene. Before the credits roll, it's revealed that the chief suspect was looking for relief from his own set of "identical bites."

Before anyone starts wondering if it's possible to "match" bug bites, consider a case from Ventura County. Entomologist Jim Webber was called in to analyze chigger bites on both a suspect in the strangulation murder of a young woman and the investigators on the scene. Both of them had been bitten all over their legs and waists.

Entomological canvassing of the region indicated that only one location—the murder site—supported the chiggers that had bitten the suspect and investigators.

Bodies provide ideal growing conditions for many insects, but forensic entomologists also get opportunities to pursue their avocation without sticking their gloved fingers into dead bodies.

Khee's baptism in forensic science sent him scurrying in, over, and under a 1953 Ford that was later determined to have been stolen goods transported across thirteen states, before finally being stopped less than six kilometers from the Canadian border. "As often is the case when cars are stolen to order, the merchandise gets handed along from driver to driver, hidden in a bunch of garages, even transported inside enclosed trucks to prevent it being seen. Each guy who handles it claims he

thought it was a legitimate moving job, and it gets very difficult to track it back to the original thief."

And bugs come into it how?

"A 1953 Ford Fairlane has a beauty grill: big, metal. Modern cars hardly have a grill at all, but one like that picks up insects like a sponge. I spent about six hours picking bugs out of that grill, out of the windshield wipers, from the treads of the tires, between the seats. Ended up with about 3,000 insects or insect parts. Insects are pretty regional. They don't, generally, migrate long distances like birds. By laying out the samples, identifying the more exotic specimens, and plotting them to a map, we could determine where the car had been geographically, sometimes to a pretty particular geography."

Mites from a damp, earthen-floored garage were found in the undercarriage and the grill—mites that were found in three other cars that all passed through the same small, one-man shipping company. It was enough for a warrant to search, and the search turned up four more stolen cars awaiting transport. The jury found the lack of any legitimate cars in the lot significant.

Coincidentally, Khee's second case didn't involve a dead body, either, though more than a car was stolen. "It was a kidnapping, and, probably because there didn't seem much to go on—entomologists aren't usually the first guys called to a kidnapping—they called me to walk through the victim's home a few days later. I didn't have much expectation. Insects come to bodies and we didn't have one. Sure didn't *want* one."

Khee and an assistant went all over the house, but there was nothing out of the ordinary. Summer had been nice and warm—good conditions for insects, except that it had been dry for weeks. The local fire indexes were through the roof, no-burn orders were in effect, bugs everywhere were actively looking for water. Khee and the investigators were leaving the scene when the assistant mentioned the flies swarming over a puddle a little farther along the street.

"I was just launching into my senior-examiner-instructing-junior-aide-mode on using water traps to collect insects in dry weather, when I noticed that not *all* the bugs were where they should be. Instead of sucking up that water, they were congregating in clumps on dry ground." It could have been something as innocuous as drips from an ice cream cone but, having just left a scene, that's not what Khee was thinking. Walking closer, he realized these flies weren't the type to congregate around food smears.

Three small groupings of flies—two in a dirt tract between the victim's house and the adjoining property, and one in the driveway of the next home—weren't evidence of any kind. But both Khee and the investigators started wondering if they might need to look closer to home.

"I couldn't tell them anything except that those flies were attracted to bodies and body fluids, usually within X number of hours or days, and that hot, dry conditions would alter the time frame. My estimate of when body fluids or blood deposited there might attract that particular fly matched closely with the time when the victim was last seen."

The victim was discovered within two hours, tied to a chair in a small shed in the neighbor's backyard. A broken nail on one hand had dripped blood into a small puddle under the chair, attracting the same flies as those Khee and his assistant had first spotted outside.

Khee's major casework since then has centered on murder and rape scenes, but he's hopeful that further developments will include "secondary use" of insect information, more cases like the drug importation work he was undertaking this spring. Inside a fume chamber, he'd been shaking out pound after pound of hash and marijuana, picking up insect remains, capturing live specimens and searching for eggs or larvae.

"Plants are the natural prey of huge groups of insects," Khee notes. "The question in this particular case is whether this is home-grown marijuana or if it's coming in from somewhere

else." He points to a brilliantly iridescent beetle. "I can guarantee that big guy didn't come from Kansas!"

EARLY FORENSIC SCIENCE

Khee's observation of bugs gathering at drops of nearly imperceptible droplets of blood harkens back to the earliest known case of a forensic investigation revolving around entomology.

It was sometime in the thirteenth century when a worker, one among many cutting crops in a field, was murdered. Conventional questioning revealed little, certainly not enough for the investigator to condemn anyone. Instead, he assembled the workers, asked them to lay their sickles on the ground and just sit. The sun shone, the day turned warm, bugs appeared to bother the workers. But only one sickle won the attention of all those bugs. Immediately confronted with the evidence—evidence that all his fellow workers could understand—the killer confessed.

Forensic Botany and Zoology

In a sense, botanical and zoological forensics form both the oldest and newest classifications of criminal studies.

A fourteenth-century account of a murder in the English countryside revolved around a single issue: who could have tracked in the cleavers discovered all over the clothes of a sixty-two-year-old woman found strangled in her bed? A cripple, she certainly hadn't been wandering the countryside where the little hooked seeds could become attached to her clothes.

As the sheriff noted to the jury, "Of her three sons, only one had the clinging nuisances on his own clothes. And only the three of them would the servants allow in their mistress's private rooms. She was alive at breakfast, dead at noon, and only they were inside the door."

Open and shut case.

Ellis Peters realized when she created her Brother Cadfael character that her crime-solving monk wouldn't be darting out

to the nearest DNA laboratory to solve his crimes. "But neither did Ms. Christie's Poirot or Miss Marple, or Conan Doyle's Holmes. A good eye for detail and a sound understanding of human nature, those are the things that solve crime even today. Especially a good eye for detail."

Peters herself must have had an observant eye, as she certainly bestowed one on Cadfael. In *The Sanctuary Sparrow*, Cadfael provides an example of how forensic botany might well have worked in early spring of the year 1140 in England. From the body of a man pulled from the river, he recovers "two or three strands of water crowfoot, cobweb fine stems with frail white flowers," "a morsel of an alder leaf," and "fox-stones . . . the commonest of its kind, and the earliest." He quickly determines that "It might be possible to find that place somewhere on the town bank—where crowfoot and alder and fox-stones all grow together." The spot, on the doorstep of the killer's home, is found in due course, as it must be in fiction, and points Cadfael and his cronies in the right direction.

Not every botany study leads to success, but it's a rare criminalist who doesn't know of at least one case that hasn't turned on botanical evidence. And some actual scenarios are as dramatic as any fiction.

Whole leaves and flowers, as easily seen by suspects as investigators, can be removed without much effort, but pollen (the microscopic reproductive cells of plants) clings to all classes of surfaces and gets mixed into common dirt, water, and even earwax.

One of the early cases (1969 is early in this field) unfolded as neatly as any novel. An Austrian man disappeared while on vacation in Vienna and the police quickly generated a likely suspect but couldn't place the subject anywhere near Vienna. Hoping the more traditional method of geological analysis might turn up something (keeping in mind that palynology, which includes pollen studies, barely existed as a forensic science) police sent the suspect's muddy boots to geologist Wilhelm Klaus for soil investigation. Klaus found soil, but he also found modern-day alder, spruce, and willow pollen, as well as fossil hickory pollen dating back twenty million years! On that information, Klaus identified one area—in Vienna—where that combination of pollens could be found. Confronted with proof that he'd been there, the killer not only confessed to the crime but brought investigators to the spot where he'd buried the body, which turned out to be well inside the area Klaus had marked.

Open and shut case.

That "cross crafting"—the tangential or overlapping nature of forensic fields, as in the Klaus case where a geologist developed pollen information—suggests that pollen or botanical study might provide another tool in much broader use. A questioned document was once declared a forgery when pollen found in the ink proved the document could not have been created in the geographical region claimed. Not surprisingly, pollen study of imported drugs and firearms often reveals the tiny grains in gun oil or coating leaves. A robbery investigator in Australia once received a strong lead when pollen was found in an otherwise useless smudge believed to be a cheek print on the surface of a safe.

Botanist Kenan Cole loves pollen. "It's so small, criminals don't see it, so they don't try to get rid of it. They'll wipe away prints, but they aren't looking for something smaller than the tip of a pin. I'll take any advantage I can get!"

Like their entomological colleagues, palynologists work regionally and apply their knowledge globally. Cole also loves planted gardens. "With plantings, you occasionally find something really unusual in an otherwise 'boring' region. If everyone in town has nothing but poplars in their yards, it's hard to differentiate one yard from another on the basis of pollen. If one chap has a eucalyptus and his neighbor has a pomegranate, we can work with that."

Just as insects have strictly regulated life cycles, so do plants. Pollination takes several forms, occurs at specific times of year, and may even be day or night specific. "Self-pollinators might produce the tiniest smidgen of pollen—it's not like it's got to go far or anything—and you'd never come in contact with it unless you actually rubbed up against the flower directly. Others scatter pollen over acres. Flowers that close at night clearly aren't spreading anything then." All of which assists botanists to limit the windows of opportunity.

To take advantage of pollen or spore evidence, sampling and collecting techniques must adhere to strict guidelines. "You can hardly walk into court and accuse everyone in the world who has marijuana pollen somewhere on their person of being involved in the drug trade. Marijuana is a wind-pollinated plant, so it makes lots of pollen. Like most wind-dispersed pollens, marijuana pollen is tiny, tiny. The reproductive strategy is to get as much pollen as possible out there, and to keep it in the air as long as possible. Marijuana flowers are three or four feet up in the air, so pollen has to stay up there, too. In a good growing season, with a nice breeze, pollen from marijuana plants can travel a hundred miles. Not everyone in contact with it is involved in growing it. You've got to know what the 'background noise' of a particular pollen in the region might be before even suggesting some association between plants and other items or persons."

Contamination in pollen studies is more likely than in other forensic fields. "Few examiners have blood about their person on arriving at work in the morning. But it's entirely plausible they'd have pollen somewhere on their person. If the sample study is to mean anything, then the examiner needs to be very aware of the means and opportunity for cross-contamination. In our lab, we handle pollen samples in hoods at differential pressure so nothing is coming into the sample area. Other policies and procedures are equally effective—as long as the samples and control exemplars were properly collected."

Keys to obtaining significant samples and exemplars include the following:

- Proper collection and packaging techniques. Each sample must be stored in air-tight containers, otherwise cross-contamination—or outright loss of the sample—can occur.
- Samples and exemplars must come from known locations and be properly dated and *timed*. Airborne exemplars collected at night likely will not contain the same constituent pollens or spores as those collected at noon.
- As spores and pollens have different drop rates (the speeds at which they fall from the air to the ground), airborne and ground soil samples should be obtained whenever possible.
- Sampling of persons or objects must include detailed documentation. The pollen in an automobile's air filters, which may be original to the car, can include pollen from every region through which the car has ever passed. Dirt containing pollens from tires may be considerably more recent.
- Spores, unlike pollen, are ready to grow once they find the right conditions. Unwary examiners could become mushroom farmers!

Entomology and the Modern Sleuth in Print

Strictly speaking, entomology is a part of the larger field of forensic zoology. But its broad application and independent development have meant that, in the real world, it's become a distinct area of expertise, while much of the rest of forensic zoology is practiced chiefly by consultants whose major work lies in non-forensic areas.

Several recent novels have made entomological findings key evidence in their plot lines. No doubt other zoological work will also worm its way into fiction—probably as real-world casework makes news. If future work maintains the high scientific standard already found in works like Kathy Reich's *Death du Jour*, readers can expect ripping stories that never leave them any room to doubt.

This brief excerpt from *Death du Jour* illustrates yet another side of forensic entomology:

"It's not new. The increase in drug-related deaths in recent years has prompted research into testing for pharmaceuticals in carrion-feeding insects. I don't have to tell you that bodies aren't always found right away, so investigators may not have the specimens they need for analysis. You know, blood, urine, or organ tissues."

"So you test for drugs in maggots?"

"You can, but we've had better luck with the puparial casings."

The fact that humans live in a world where they, animals, and plants interact on a fairly regular basis should go without saying, but cases like this one illustrate that rather nicely. And there's a happy ending!

In 1993, a woman was assaulted in a public park. Fortunately, she managed to escape with little physical harm beyond a purple shiner and a series of scratches across her neck. She did bring away physical evidence, though. A cocklebur, with a family

of insects in residence and several long, white hairs attached to it, was found attached to her clothes. As it certainly hadn't been there when she left to walk, and there were none to be found at the place of her attack, investigators realized the assailant must have transferred it to his victim via the blanket he'd attempted to throw over her head.

Luckily for her, the local authority had officers and investigators capable of recognizing the significance of the evidence they had in their hands, and they quickly called in several consultants. The hairs, attributable to the fox once very popular for collars, were identified by Calvin Dean. Marie Coule, examining the cocklebur and determining it to have come from the area immediately around an abandoned campground, handed along the insects to Joseph Harvey, who identified the species. When the campground was investigated, they found someone had broken into one of the cabins and was living there. He'd also ransacked the other buildings, including the owner's cabin, and had taken the blanket from a trunk that also held a woman's coat, circa 1930. The coat had a long fox collar, a collar that had apparently first attracted the attention of the bugs that had made themselves at home in the cocklebur. Anyone hear echoes of Cadfael's "where all three . . . come together?"

THREE

WORKING THE SCENE OF THE BODY HUMAN

So much evidence is found on and inside the human body that it constitutes a crime scene unto itself. After undergoing preliminary examination at the scene, a body is released into the hands of coroners or medical examiners, becoming the landscape that yet another rank of investigators will scrutinize intimately—right down to its basic molecules, in fact.

Blood Factors and the DNA Fingerprint

Blood's individual qualities were recognized years ago when doctors realized some transfusions were successful while others were immediately fatal. From that observation came the ABO typing system and the first understanding of the Rhesus factor.

While there are extremely rare or exotic blood types, most people can be classified into the A, B, O, or AB blood types. The fact that an AB type exists at all told early investigators that every individual actually carries *two* alleles, or traits that determine blood type—one inherited from each parent. Further studies proved that if each parent contributed an O allele, the child would be type O, but if one parent contributed an A and the other an O, the A dominated. Likewise, if one parent contributed a B and one an O, then the B dominated.

A and B do not dominate one another, though, so if one parent contributes an A and the other a B, the child displays both traits and is AB. Obviously, having no A or B factors, two O parents can only have OO children (usually just called O). Two AB parents can never have an O child, as neither of them has an O allele to pass along. Two parents of types AO and BO could combine their O's to produce an O child, their A and B to produce an AB child, their A and O to produce an AO child (called A), or their BO to produce a BO child (called B).

In paternity cases, studies of family history can often reveal if an individual is AA—receiving an A type from both parents—or an AO—receiving an A from one and an O from the other. But for the practical purpose of receiving or donating blood, it doesn't make a difference and it isn't possible to separate AO and AA from one another by any sort of testing.

Another factor present in some blood is the rhesus factor. It is called rhesus because it was first identified in rhesus monkeys. A person that is A+ has type-A blood *with* the rhesus factor. A person with A− has type-A blood but *doesn't* have the rhesus factor.

There's a common misconception that there are two strains of the rhesus factor—one positive and one negative. That's not true. You either have it or you don't, but you can't have a negative version.

This factor can be important during pregnancy if the mother is Rh-negative—that is, if the mother does *not* have the rhesus factor. Children conceived with an Rh-positive husband could receive the rhesus factor from him. The mother's body, which doesn't have it, perceives the new factor as an invader, like a cold virus, and starts building defenses against it. This usually happens during a first pregnancy with an Rh-positive child. If a second pregnancy with an Rh-positive child occurs, the defenses developed during the first pregnancy swing into action and serious complications can result, which is why physicians urge blood testing for both parents when a pregnancy is discovered. Now that the rhesus factor is understood, treatments can allow second and subsequent pregnancies to proceed normally.

With just ABO typing and rhesus testing, there were eight possible blood types: O−, O+, AB−, AB+, A−, A+, B−, and B+. For *excluding* suspects, this was occasionally sufficient. If the blood under a victim's nails was B− and the suspect's blood type was O−, there was no match and the suspect was cleared. Problems arose when the blood typing *did* match. Obviously, law enforcement couldn't arrest everyone with B− blood!

Statisticians knew that some blood types were more common, or more common in certain ethnic groups than in others, and tried to bring mathematics to blood typing. But juries just weren't having it. Blood typing remained good circumstantial evidence, but it wasn't going to convict anyone on its own. Unlike fingerprints, which were unique to an individual, blood types could be shared by millions of people. Other biological clues, like the Barr bodies found only in the cells of females, might help narrow the pool of suspects, but even half a million people is a pretty big pool.

For that kind of individualization, law enforcement had to wait until DNA analysis—the study of deoxyribonucleic acid—advanced. DNA, the physical material that we inherit from our parents when that one sperm finds that one egg, is absolutely individual—with one exception. Identical twins form from one fertilized egg, so **their** DNA is identical.

While fingerprinting is still the only surefire way to separate identical twins, DNA testing provides its own advantages:

- Every cell in an individual's body contains identical DNA. Fingerprints come only from fingers, but DNA can be found in blood, in urine, in feces, in saliva, in some hair, in the shed skin cells found in a facecloth or toothbrush—even in the sweatband of a hat! A suspect doesn't have to bleed at the scene to leave DNA. Semen at rape scenes, saliva on the envelope of a ransom note, skin cells scraped onto a rope while tying a victim—all provide the opportunity for collection and analysis.

- DNA can survive much longer than a fingerprint. While some few prints have been collected years after being made, DNA analysis has been performed on Egyptian mummies! Granted, not too many mummies are going to be arrested, but the example helps illustrate DNA's power as a means of identification. DNA can provide closure to a victim's family years later by making it possible to identify bodies that might have been buried just a decade ago as John or Jane Does.

- DNA can indicate familial relationships. Though it's been theorized that prints in family members might have similarities, it's unproven. But because DNA is inherited from two parents, a significant number of matches in a sample can point to a "first-degree" relative—a mother, father, or sibling. In cases where groups are involved in crimes, this is important evidence. A Philippine case involved a murder conducted by two individuals. One was identified by an eyewitness, but the second man was not. DNA from spit at the scene didn't belong to the man already identified, but had so many points in common with his that investigators suggested a sibling as the second individual. Brought in on that evidence, the brother confessed.

- DNA evidence doesn't combine. Blood evidence at a scene frequently comes from more than one individual—either the attacker and the attacked or a number of victims. If one hypothetical victim is type A and another is type B, a combined sample of their blood might suggest an AB individual. With DNA, the traits of both victims *would* be found in the sample. But with samples from the victims themselves available for comparison, it would be possible to prove that it's a combined sample possible from *only* these two individuals. It makes determining who was at a specific point in the scene possible. Blood evidence could have suggested the inclusion of completely fictitious individuals!

With so much at stake, collecting DNA evidence at the scene, as well as from both victims and suspects, is imperative. But it's not always easy. Collecting and preserving quality DNA samples requires that investigators first determine where DNA evidence might be found, and then ensure that it is collected without any contamination occurring to what may well be a tiny sample.

Common sense suggests beds as likely places for semen samples and telephone receivers and cigarette butts as possible sites for saliva, but investigators know all too well that common sense won't account for every possibility.

Having worked as a criminalist for eighteen years, Paresh Mosley has found DNA in some pretty bizarre locations.

"Behavior, even in humans, must have pretty basic underpinnings somewhere. After working a bunch of crimes that took us to cabins and other woodsy locations, we eventually came to the conclusion that, if you put ten guys in the woods and they've got to take a leak, nine of those ten guys will prefer to pee on a tree trunk. Now, I'm sure that there's no manual to manhood that says you should pee on a tree, and I'd like to think we're not 'marking' our territory, but you shine a source light around a campsite and I'll guarantee you there'll be DNA from urine on some tree somewhere."

Common sense, good observation skills, and a knowledge of human behavior lead investigators to the evidence—Ms. Christie and Ms. Peters would be pleased.

"Sometimes you get thrown. Working a really grisly rape-murder, we expected to find any semen on the body either around the victim's mouth or in the genito-anal regions. That would be typical. When we didn't, we started assuming the suspect used a condom. On postmortem exam, they did find semen, but it was in one of the numerous stab wounds on the victim's torso. Thing like that, well, you just can't conceive of it until you've been confronted with it."

More typical searches turn up blood between wooden

floorboards, in drains where suspects have attempted to clean up after themselves, inside ski masks where hair samples may contain the valuable roots where DNA is found, or, on human beings, in bite marks or under fingernails.

Before it can be collected, it has to be found. And if one of DNA's advantages to the investigator is that small samples are still useful, then one of its disadvantages is how hard to spot such tiny deposits can be. Two of the aids investigators may use are chemicals like luminol and alternative light sources like UV. Many biological samples fluoresce naturally in certain wavelengths of light. Semen, blood, and amniotic fluid are amenable to this method. If it's not possible to use ALS, a spray of luminol (which reacts with the iron in blood and any other blood-containing biological specimens) proves valuable.

In the United States, the National Institute of Justice has a checklist of items investigators should consider in determining where DNA evidence might be found and suggests the collection of these:

- fingernails or fingernail parings
- tissues, paper towels, napkins, cotton swabs, ear swabs (bag *everything* in a bathroom wastebasket)
- toothpicks, cigarette butts, straws, anything else that might have been in contact with the mouth, like cellular phones
- blankets, pillows, sheets, mattresses, dirty laundry
- head gear of any type
- eyeglasses, contact lenses
- used stamps, envelopes
- tapes, ropes, cords, anything else used as ligatures
- used condoms
- bullets that have passed through bodies

Clearly, many of these sources will be submitted for more than DNA examination, so it's essential that prioritization and joint handling procedures be agreed upon by all investigation members to avoid the loss of evidence.

WANTED–FORENSIC DNA ANALYST/ DNA TECHNICAL LEADER

IDENTIGENE® is seeking applicants for Forensic DNA Analyst/DNA Technical Leader positions.

Qualifications include a bachelor's or master's degree in biology, chemistry, or forensic science; a minimum of six months forensic DNA casework experience to qualify for the Forensic Analyst position; a minimum of three years forensic casework experience to qualify for the Technical Leader position. Must have completed course work covering the subject areas of biochemistry, genetics, and molecular biology. In addition, course work and/or training in statistics and/or population genetics is necessary. Experience handling a broad range of forensic samples, experience giving expert witness testimony, and familiarity with the ABI Prism™ 377 Automated DNA Sequencer are desirable.

These positions report to the Forensic DNA Laboratory Director.

Responsibilities include performing DNA profiling on biological samples; performing test result interpretation; giving expert witness testimony for criminal casework; performing internal validation studies for automated STR systems; and participating in training.

Salary: $35,000–$65,000/annual.

Since DNA is found everywhere, extraordinary efforts must be made to isolate samples one from the other and to prevent investigators from bringing in inadvertent samples. Investigators are encouraged to follow these guidelines:

- Bring lots of gloves and change them often.
- Where possible, use disposable tools. Holding individual swabs with disposable craft sticks prevents transfer of DNA evidence from one sample to the next, as might be the case if reusable items like metal forceps are used.

- Avoid contact between gloved hands and face or hair. Even unconscious habits, like pushing eyeglasses up on the bridge of the nose, can introduce investigator DNA.
- Don't touch surfaces except with collection material, such as swabs.
- Investigators can avoid spraying their own saliva-based DNA over a scene by refraining from talking, sneezing, or coughing. A mask makes this much easier to control.
- Most DNA evidence, being biological evidence, is damp. It must be air-dried, then packaged in non-plastic containers to prevent mold or bacteria from creating a whole different kind of "science experiment" inside the containers.

All evidence collection should be undertaken in as timely a manner as possible, especially if biological evidence is involved. Humidity, direct sunlight, rain, and the growth of bacteria can all degrade samples beyond the point of recovery. And removing the evidence to the inside of an evidence van sitting in bright sunlight doesn't solve the problem, it just bakes the evidence. So air-conditioned or cooled containers are essential equipment.

The flip side of the contamination issue is that DNA sampling, which brings investigators into proximity with any number of bodily fluids, requires special precautions to ensure the *investigator* isn't contaminated by the *samples*. Biohazards like hepatitis and HIV are real dangers to investigators and laboratory personnel who handle bio-wastes and bio-fluids, so all samples must be treated as infectious until proven otherwise.

DNA permeates every cell of the human body, but investigators aren't limited to direct sampling of body cells or body fluids. Blood recovered from flea, cockroach, and mosquito guts can and has been successfully submitted for ABO and DNA typing. Human blood in cockroach feces turns up under luminol testing and has actually confused the blood splatter analysis when fecal matter deposited around a room (even on the ceilings!) has been mistaken for dried blood droplets.

NON-HUMAN DNA

Every living thing contains DNA. Plants, cats, mushrooms—everything. Not surprisingly, criminal encounters that allow the transfer of human DNA from person to person also allow the transfer of non-human DNA.

One of the first cases linking two people through non-human DNA was investigated by the Royal Canadian Mounted Police (RCMP). While investigating a death in Prince Edward Island, examiners recovered two white hairs, which were at first thought to be those of the victim's ex-husband. They weren't. They were cat hairs. Not having a unit that dealt with cat DNA, the RCMP sent the hairs, as well as a blood sample from the ex-husband's white cat, to the National Cancer Institute's Cat Genome Project in Maryland. The lab there confirmed that both blood and fur came from the same cat, and the ex-husband was convicted of murder.

As this goes to press, DNA studies specific to literally hundreds of plant and animal species are underway. It may soon be possible to analyze and identify to a particular source every single living creature a body has contacted.

With a good, uncontaminated sample that hasn't been baked or grown moldy, lab personnel can decide on a typing technique. The earliest DNA typing was the RFLP (restriction fragment length polymorphisms) method. In this technique, DNA extracted from the sample is cut into fragments by chemical "scissors," which separate the long DNA strands at specific spots between any two of the four proteins that make up DNA. In chemical shorthand, the four proteins are A, T, C, and G. Because everyone's DNA is different, the length between one person's A's and T's or C's and G's is different from the distances between another person's.

Suppose we have two suspects and a pair of chemical scissors that will snip every C to G bond in a string of DNA proteins.

In suspect #1, CGATTAGCGAGCT becomes C GATTAG C GAG CT.

In suspect #2, TTCGTATATATACG becomes TTC GTATA-TATAC G.

There are some very short pieces, some middle-length pieces, and some very long pieces.

Next, a gel (a suspension that can carry an electrical charge) is set up. (This part of the process is called gel electrophoresis.) DNA from each person, a control sample, and the sample from the unknown are set into different lanes of the gel so the material doesn't get mixed. Then current is run through the gel. When that happens, the fragments begin to line up. Lighter pieces (the shorter ones) travel much farther than the heavier (longer) pieces and, when the process is complete and everything is at rest, the fragments will have formed into distinct bands based on their weight. These bands become visible when they're tagged and x-rayed. It's usually these films that are seen in court.

If two samples come from the same person, they'll break along the same lines and come to rest in the same places. That makes a match. If they aren't in the same place, there's no match. If there are considerable similarities but no match, it's possible the samples are from related individuals.

The RFLP technique is a detailed, accurate, and precise way of identifying individual people. On the downside, RFLP is slow. It can take anywhere from three weeks to three months to get results. The lab work alone takes nearly a month and, because it ties up lab facilities, there's frequently a backlog of work. Not helpful if you're a law enforcement officer who needs a lead. And certainly not helpful if you're the accused waiting for these results to clear you! On top of that, RFLP requires a **lot** of **fresh** sample. Large blood splatters, hair samples of up to twenty-five hairs, or large quantities of saliva are needed for each test. A semen stain the size of a large thumbnail may not contain enough DNA for testing. If you only have one sample, this method might take up all the available raw material.

PCR (polymerase chain reaction) methods improved on the

RFLP technique by replicating the DNA present in a tiny sample until there was enough of it to type, making even minuscule samples significant. It works on the principle that DNA replicates itself naturally each time a cell divides. PCR creates the same situation chemically by "unzipping" the DNA molecule into its two halves. Because the exposed molecule ends on each half can only attach to their opposite chemical numbers—supplied in raw form by the technician—the created half is identical to the original. When both halves have grown new halves, the amount of identical DNA has been effectively doubled. And the process can be repeated by unzipping the new bits and allowing the chemical reactions that create new halves to recur as often as is needed to obtain a large enough sample. From that sample, it's possible to compare evidence based on the locations of specific markers on just a single stretch of DNA. The results from PCR testing of this type are seen as a series of dots. If the dots are darker than the control dot, or C dot, it's a match; if not, it's not. PCR typing is a list of the reactive points—for example, "1.3, 4."

The advantages of PCR are many. To begin with, it's fast. Even the longest testing period should take less than a week, and it's frequently faster than that. Some preliminary results can come back in forty-eight hours. The whole process of duplication occurs in a single vial that contains the sample to be copied, lots of raw material for the natural duplication process, and some "primer" (the chemical that starts the duplication reaction). The vial is first heated to about 160°F for thirty seconds. This allows the two halves to unzip. When it is cooled to about 100°F, the primer kicks in and, in less than half a minute, the process of duplication begins. The vial is reheated to about 140°F and duplication continues until both halves of the strand have been copied. The whole process takes about two minutes! And it's repeatable. In less than four hours, it's possible to multiply a sample by 5,000 percent.

Also, PCR testing is much less expensive than RFLP methods, making it possible for more case evidence to get into the system.

PCR can also use "degraded" evidence—in other words, older samples recovered at a secondary scene located some time later, even up to decades after the primary scene has been worked and—can still produce significant results. Tiny samples—a single hair, blood or semen stains smaller than the ball of a ballpoint pen—are sufficient.

Unfortunately, PCR samples are at higher risk for contamination than are RFLP samples. The copying process can't distinguish between DNA left by a suspect and DNA from a criminalist who breathed a little too heavily—consequently, all DNA present is copied. From a purely statistical perspective, because of the narrower region of study, the results may not be as detailed as RFLP. One in millions, or even billions, are terms associated with RFLP; with PCR, the numbers are more like 1 in 10,000 or 100,000—impressive, but not of the same order.

ELEANOR CRUISE

"DNA studies in forensics used to be confined to human DNA," notes Eleanor Cruise, as she carefully manipulates tubes in a centrifuge machine and sets the thing spinning once again. "But today, it might be cat DNA, plant DNA, almost anything. Whatever trace the people on the scene can turn up."

In an environment most would already consider borderline science fiction, Eleanor Cruise is a lady-in-waiting to even more incredible feats of DNA wizardry.

"A few years ago, a case in Lafayette centered around whether or not it was possible to trace mutations of the HIV virus the way we trace paternity in human genetic studies. A doctor was accused of—and eventually convicted for—injecting his ex-lover with the blood of one of his patients who had AIDS and another with hepatitis C. A researcher there was asked to discover if the virus floating in the victim's blood was the same as—or closely related, as everyone realizes these viruses mutate rapidly; that's what makes it so difficult to treat in the first place—the doctor's patient."

Pretty high-tech work.

"Yes, it was, even by today's standards. And while I wasn't sure exactly what was or wasn't proven in court on that occasion, I could certainly see where similar work might lead in the future. Theoretically, it opened the door to proving contact in any number of cases where disease was passed from one person to another."

So, a rapist and his victim might share the common cold, instead of the usual bodily fluids, and that could be the basis of a case?

"Theoretically, yes."

Which is why Cruise is participating in a think tank study—to see how DNA studies, and other branches of genetic research, might aid investigators of the future.

"We aren't really trying to solve a particular crime as much as we are trying to discover new parameters—based in good science and practical on the level that could be implemented by police and medical personnel—whereby we could recover evidence that hasn't even been contemplated today.

"For example, several years ago now, a researcher managed to match blood found in a flea's stomach to a victim. As the flea was found in the suspect murderer's clothing, there was an immediate, undeniable link between them. We've matched pollen from crime scenes to pollens found on stolen vehicles. This is really just the next step—to match the DNA characteristics of bacteria and virus samples to one another so we can determine if two individuals were in the same environment, or in close enough contact with one another to pass microorganisms back and forth."

But is that practical, with the amount of mutation common in some organisms?

"It's a problem, certainly, but not insurmountable in all cases. In fact, there are even some cases where a *lack* of mutation is more problematic. Some bacteria, for example, are incredibly stable generation to generation, with almost no genetic drift at all. If that bacteria is commonly found in many areas, there's nothing distinct enough about it to positively link subject A with subject B. So, yes, there are all sorts of limitations to what may be accomplished. Which is why projects like this think tank are so important. If we can determine rates of mutation and relative stability in various strains of micro-organisms, then we can look specifically for those that give us the best chance of including or eliminating suspects."

But none of this work has been taken into the courtroom as yet?

"No, not yet. But it will be. Shortly after the discovery of DNA, a researcher wrote a paper and presented it in Berne speculating that, by the year 2000, every individual on the planet would be iden-tified by a DNA-marker test done as routinely at birth as the Apgar has been in recent years. We didn't quite take science that far that fast, but it's coming. The United Kingdom has DNA data banks on sexual offenders, and that trend is spreading."

When do you see germs becoming witnesses?

"Within the next eighteen months. Certainly within the next twenty-four."

So, criminals will have to be very sure they aren't sniffling when they take out the local convenience store?

"I think the point of all forensic work is to simply make it so diffi-cult to commit a crime that people will simply not bother anymore. At least, that's what we're hoping!"

DNA analysis is a quickly changing field. Just when PCR was becoming generally understood, its values and differences from RFLP filtering down to street-level investigators, new tests—the STR (short tandem repeat) process in particular—began yielding significant results. STRs are newcomers in court, but have been used extensively in non-forensic identifications. Vic-tims of TWA Flight 800's crash in the Everglades were identified by this method, as were those who died in the Branch Davidian fire outside Waco. And it seems likely that the victims of the September 11 assaults may be identified by the same methodol-ogy. As might be inferred from these incidents, STR techniques can function with severely degraded DNA.

The small amount of DNA available is multiplied using the PCR technique, but the raw material used in the duplication process has already been tagged with fluorescent dyes. When the gel electrophoresis process is complete, the bands can be read with lasers instead of X-rays, and the analysis of the locations can be read by a computer. Multiple samples can be run at once, and the results are a considerably more readable graph—the electropherogram—instead of X-ray films. STR also offers the potential testing of hundreds of well-documented loci along the DNA strand, making it possible to achieve a high level of discrimination—and putting the science back in the one-in-a-million club. Additionally, two different methods of visualization are available—silver blotting on gel or fluorescent tagging. Silver is cheap in the field of DNA analysis, while fluorescent tagging may someday make real-time analysis possible. Cheaper, faster, flexible, and computer friendly—all a forensic scientist could want.

WANTED—
FORENSIC CHEMIST-BIOLOGY (DNA) SPECIALIST

The Acadiana Criminalistics Laboratory (New Iberia, LA) is seeking applicants for the position of Forensic Chemist-Biology (DNA) Specialist. An M.S. degree in biology, biochemistry, molecular biology, genetics, forensic science, or related field, with three years of experience in biology (DNA) is highly desirable. A B.S. degree in biology, biochemistry, molecular biology, genetics, forensic science, or a related field, with one year of experience in biology (DNA) will be considered. A minimum of one year of casework experience, including testimony as an expert witness in the area of STR PCR DNA typing, and familiarity with the ABI 310 CE DNA Sequencer are highly desirable.

Applicant must possess knowledge of CODIS, DAB, and ASCLD/LAB regulations, have a strong background in quality assurance systems within crime laboratories, possess knowledge of laws governing rules of evidence and courtroom procedures, be able to communicate effectively orally and in writing, be able to organize and coordinate laboratory activities, be able to effectively supervise laboratory support staff, be able to use computer applications as they relate to laboratory procedures, and possess a valid driver's license.

Responsibilities include conducting DNA casework analysis using PCR STR ABI 310 CE technology, acting as liaison with law enforcement officers, selecting probative exhibit materials, identifying and comparing biological material, writing case reports, and providing court testimony as an expert witness.

Potential employees are subject to background clearance, pre-employment and random drug screening. Travel and overtime for court, lecturing, crime scene attendance, and training responsibilities is required.

Salary with M.S. degree: $45,834–$87,259/annual; salary with B.S. degree: $40,081–$81,505/annual.

DNA analysis (frequently called DNA fingerprinting) does share one other characteristic with the more established science—its results can be coded for storage and retrieval in a database. STRs easiest of all. The genetic equivalent of AFIS is CODIS (the Combined DNA Index System) and, since 1999–2000, most labs nationwide in the United States have begun testing for the same thirteen STR points, meaning that all results can be shared among agencies.

IT'S A GIRL THING

Nuclear DNA (the DNA used in RFLP, PCR, and STR analysis) comes from both parents, but in an egg about to be fertilized, there's already a considerable amount of a different sort of DNA floating about. It's called mitochondrial DNA (mtDNA) and is passed from mother to child. As it can only identify the mother and most paternity testing is designed to determine who the father is, it's never been a focus for practical development.

Its unusual features are now coming into their own. Because a mother and her children share identical mtDNA, it's possible for samples from an unknown to be compared to the mother in the absence of her child, and vice versa. And mtDNA can be obtained from bones, teeth, and hair—traditionally difficult subjects for DNA. Like PCR and STR typing, mtDNA typing works with degraded and limited samples.

Forensic Odontology

Forensic odontology, or forensic dentistry, provides an opportunity for identification when friction ridges or regular DNA fail. In fact, it may be preferred to DNA in many jurisdictions where time and distance make DNA analysis slow or expensive. Because teeth can survive when other tissue doesn't, it can be the *only* evidence available for identity testing. Additionally, that tough outer covering can protect mtDNA inside the tooth, in the pulp, providing a reservoir of genetic material.

"Teeth are terrific," asserts dentist James Berry, "Give us one tooth and we just might be able to make an identification. Give us a mouthful, and we can often match suspects to victims, unknowns to names—even evidence items to victims and suspects. They last almost forever. They're well-documented in most cases. They're about the toughest biological evidence available."

The job of a forensic odontologist breaks down into two main activities—matching teeth to individual records and matching teeth to bite marks. And, though bite mark evidence wasn't introduced until the 1970's, odontology has been identifying people for at least 2,000 years. In America, Paul Revere, a dentist, helped identify the dead of the Revolutionary War based on dental evidence. Dental records identified Adolf Hitler and Eva Braun. They even helped to secure a conviction of Ted Bundy, who liked to bite his victims.

Modern applications of odontology include more than the visual examinations Paul Revere used. Radiology (X-ray and other techniques) allows otherwise hidden details to be visualized and captured as an exhibit. The SEM (scanning electron microscope)

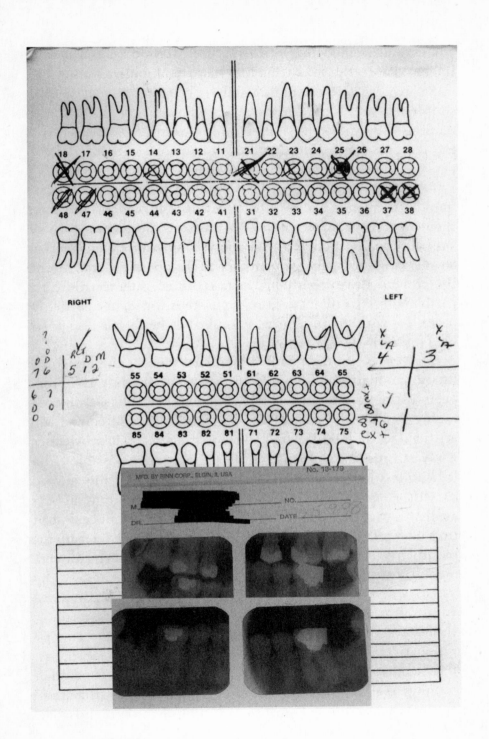

and computer modeling permit a bite mark to be matched to the mold of teeth still sitting in a suspect's mouth thousands of miles away. There exists no database of teeth, but given a description based on the Universal System (charts of all thirty-two teeth used by all dentists), thousands of records in offices across the country can be checked for similarities, without the evidence ever being shipped to any of them.

Two different systems age teeth quite accurately. Gustafson's method identifies six signs of wear on adult teeth and rates them numerically per tooth. The Lamendin system works on the other end of the teeth—the roots. Transparency in the roots changes with age—lightening begins at the tips and works toward the crowns. Baby teeth don't have roots as adult teeth do, but particular teeth fall out at fairly consistent times, in a particular order. So, with a mix of primary and adult teeth, it's also possible to narrow the age range.

In all cases, dental records—the usual tool for comparison of unknown dentition—contain a description of every tooth and every extraction, as well as notes on caps or fillings, twisted or malformed teeth, unusual root formation, and, of course, any artificial teeth or bridgework added to the mouth. In most cases, X-ray records can be found, as well; these make possible radiological testing on suspect teeth—even without the suspect present. In some cases, molds of teeth or gums may also be available. Molding is commonly completed in the process of fitting odontological aids. Pictures of the interior of a patient's mouth, available in many offices now, can be converted to 3-D models if sufficient views exist.

PERCY MICHAELS

"In any field, there are cases that stand out or become references for later work," says Percy Michaels, as he carefully examines the photo of a bite mark. "In our field, in the United States, the Bundy case is like that because it was one of the first highly publicized cases to truly turn on bite-mark evidence. But there are others that, at least to those in the field, were considerably more eye-catching. One of my personal favorites didn't actually involve a crime at all, at least not in the traditional sense. But it took odontology to prove it."

Propped on a light box are two sets of skull X-rays–full frontal views which, to my eye, appear absolutely identical.

"Good call. These are the opening shots in a series of films taken of Jasmine and Jacinth, twenty-two-year-old identical twins." Sliding one image over the other, he proves that every point matches. "And they're as difficult to distinguish from one another in the flesh, as well." A long pause, and then, "At least they *were*. Jacinth, the older by a few minutes, died last year of AIDS-related complications. Before her death, however, the two women found themselves in a rather convoluted forensic mess.

"A young man of their mutual acquaintance, after learning Jacinth was HIV-positive, presented himself at his local jail, hysterically demanding she be arrested for deliberately infecting him with the virus. Once the attending officers got him calmed down enough to explain himself more clearly, it became obvious that, at some point, he and both twins had participated in a mutual sexual encounter. All the normal barrier protection was used, but at some point during the three-way event, one of the girls had bitten him firmly enough to break the skin and, though he couldn't know at this point if he was actually infected–nor could he even name with certainty which of the girls had bitten him–he was determined to have them both detained and charged with attempted murder."

Could that charge hold up?

"It was iffy at the time, but similar cases were being looked at in a number of jurisdictions. And if Jacinth had indeed bitten him, knowing she was HIV-positive at the time, there was a very good chance a charge of negligent homicide or depraved indifference could certainly have held up."

All of which presupposes it was Jacinth and not her sister who bit him?

"Exactly. And not surprisingly after a sexual liaison that all three admitted was mutually rough—the women either didn't know, or didn't wish to know, who had caused that particular bite."

Michaels drops the skull shots from the box and tucks much smaller dental X-rays under the clips. At first glance, they appear as identical as the skull series. But before long, he has pointed out seven points of difference. "Unlike DNA, which is identical in identical twins, teeth and the resulting bite marks are often as different as those of unrelated individuals."

But what were you supposed to compare these teeth to? Surely, after all the intervening time, the man's bite would have healed?

"We do get lucky sometimes. Once, I had a case where an emergency room nurse was on the ball enough to actually snap pictures of emerging bruises from a rape victim who'd been bitten, but that wasn't the situation in this instance. While the bite, on the man's left buttock, was deep enough to have perhaps warranted medical attention, he was reluctant to seek help for such a potentially embarrassing injury. So we had to call in a forensic photographer who could do UV photographs, which can peek beneath the skin and show us underlying tissue damage even months after the surface has apparently healed. And that's when we got lucky."

UV pictures of the bite seem almost magical. Where nothing but smooth skin is apparent in the normal photos, the round ring of a bite formed by both upper and lower sets of teeth nearly two months ago is clearly visible on the special photographs.

"With this image and impression marks from both sisters, it took us just a few hours to assure the man that it was absolutely impossible for him to have contracted the HIV virus through that particular bite. It was, without doubt, made by Jasmine."

For bite marks, forensic odontologists and examiners prefer to take a series of photographs over whatever period of time is available. The marks can change considerably in appearance on living individuals as bruising continues to develop. Even on the dead, lividity or blanching may change the appearance

of the bite mark; many details become visible only after the blood has settled. Examiners employ specific descriptions to convey to one another both the type of bite and the depth of the impression:

- **Hemorrhage**—small bleeding spot
- **Abrasion**—non-damaging mark on skin
- **Contusion**—ruptured blood vessel, bruise
- **Laceration**—punctured or torn skin
- **Incision**—neat puncture of skin
- **Avulsion**—removal of skin
- **Artifact**—bitten-off piece of the body

These are the degrees of impression:

- **Clear**—caused by significant pressure
- **Obvious**—first-degree pressure
- **Noticeable**—violent pressure
- **Lacerated**—violently torn from body

Because of changes that in the body—and the skin in particular—it is important to get the odontologist and the bite mark in the vicinity of one another as quickly as possible.

"I was called in about seventy-two hours after a murder in 1990," says Berry. "Took one look at the mark and, after doing what I could do at that point, took myself home and wrote a letter to the chief medical examiner, with photos of other scenes where we managed to get there early, to show him exactly the opportunities we'd missed this time. I arrived at the next scene with a bite-marked body *before* the ME—the CME called *me* first. I guess you'd say we'd come to an understanding.

"It wasn't sloppy or anything else. He simply didn't know because we'd never had bite mark evidence in our region before. The natural assumption is that bite marks are the evidence least likely to change over time, the least likely to be lost. That a picture can save them and that's that. Unfortunately, molds cast later in the process aren't nearly as valuable as those cast earlier. After a bit, skin 'slides' on corpses. I've seen people slip the skin off a victim's hands and use them like gloves to make prints! The tissue under skin can be as important as the skin itself, but once the two become separate, well, we lose detail. In some cases, the only way to capture that detail is by excising the bite itself, skin and tissue." [The method mentioned is known as the Dorian method.]

The chance to collect saliva for DNA testing exists for only a short time in bite marks. "Someone who is a forensic odontologist will remember to swab that mark. Someone who isn't may not think of it in time."

BITE ME!

Never narrow the field too early. While 99 percent of the matches a forensic odontologist makes between teeth and bite marks feature the suspect's teeth and the victim's bite mark, the other 1 percent are cases of *defensive bite marks*.

Lionel Messervier, faced with a suspect who'd bitten a victim three towns away, was understandably startled by the violent bite mark found on his *suspect's* wrist, but he simply went about his business, processing the unexpected bite mark and collecting the dental impression he'd been sent for.

Investigators weren't impressed by claims that "She bit me first!" when Lionel returned a match on the suspect's teeth—to both bites!

"I was ready to consider the possibility that she'd bitten him in self-defense. God knows she'd taken a horrible beating. I'd have bitten him if I were in her shoes. But that the guy bit *himself* to somehow make it her fault? Too weird."

Complicating the process of bite mark recovery is the likelihood of several overlapping bite marks, the confusion between upper and lower arches, and the possibility of missing tissue. The bite mark itself is of no use in identifying a suspect if there are no suspect teeth with which to compare the mark. (It is possible, however, to connect separate crimes based on bite marks on different victims.) Collecting exemplar bite marks from suspects is best accomplished by an odontologist. "You can't 'fake' dental impressions," Berry confirms. "But deliberately misaligning the uppers and lowers can create impressions that don't *look* right to a jury, so it's always a good idea to make sure the impressions are collected by someone who knows all the tricks."

Photography in bite mark evidence collection needs specific handling. Color film records differently from black and white. The UV photography technique brings out more detail at the tips of tooth penetration. The IR range emphasizes subsurface damage caused by bruising and bleeding.

New graphics programs help to trace those marks on photographs, making overlays used to compare teeth and bite marks more accurate.

Old this forensic technique may be—but it sure isn't stagnant!

Forensic Anthropology

The science of forensic anthropology includes archeological excavation; examination of hair, insects, plant materials and footprints, determination of elapsed time since death; facial reproduction; photographic superimposition; detection of anatomical variants; and analysis of past injury and medical treatment. However, in practice, forensic anthropologists primarily help to identify a decedent based on the available evidence.

FBI Law Enforcement Bulletin, July 1990,
Robert W. Mann and Douglas H. Ubelaker

Many forensic anthropologists include at least some aspects of odontology in their field of expertise, but it is the bones that primarily concern people like Wally Schier, who "came to the work accidentally," but "has since spent about seventeen years primarily in the forensic field." That's true of many of those currently practicing. "About half the anthropologists I've met at conferences have stories similar to mine—getting called in because they worked at the handiest university, getting the bug for the work, taking on forensic specialization. Most of us have some archeological training and, really, that's not much different from working a crime scene. We graph the area at digs and record the exact position of *everything*. We've already had years of training in finding minutiae, including the minutiae on and of human bodies. We already ask many of the same questions that a forensic investigation addresses, so it isn't a huge jump in thinking."

Which questions do forensic anthropologists ask? And how are they answered? An entire degree program only begins to tell a body's story, but some of the basic obtainable information includes the following:

Age:
Bones change throughout an individual's life. Wrist bones, for example, show continuing calcification until the early teens. Most parents know that the fontanelles, the "soft spots" on a baby's head, close slowly over the first year or so of life, but may not realize that the other skull sutures–the joints–continue to close throughout life. The smoother the skull and joints, the older the skull. Joint wear throughout the body, particularly of the spine, is also helpful in determining age.

Sex:
A number of skeletal differences exist between men and women, in the bottom of the skull, the forehead, the jaw, and (as is commonly mentioned in both fictional and real-world

anthropology reports) the pelvis. In women, the pelvis is lower and wider. From this set of bones, it's also possible to determine whether a woman has given birth.

Height/Basic Body Type:

From measurements of the long bones, especially those in the legs, it's possible to estimate the individual's height. Weight and muscularity can be determined by the size, position, and type of muscle attachments to the bones. This might also indicate the type of activity with which the person was involved.

Handedness:

Left- or right-handedness is also easily determined by muscle attachment; there is generally more muscle attachment on the dominant side.

Ethnicity/Racial Type:

In our increasingly multiethnic and interracial society, many anthropologists do not feel particularly confident making judgments along these lines, but some facial characteristics allow the examiner to narrow the field in identity work. For example, a Caucasian individual typically has a narrow face, prominent chin, and high nose, and the Innu of northern Canada display particular dental patterns, while the Asian facial plane has forward-projecting cheekbones. Other clues, such as any hair associated with the skull, significantly increase the opportunity to suggest racial type.

Death Time Frame:

Within the range of error caused by environmental conditions, it can be possible to estimate the time that has passed since death.

On the broader scale, violence leaves its mark on many long-lasting tissues, including bone and cartilage, and the anthropologist offers expert advice on which of these disfigurements are forensically significant. For instance, gnaw marks from rodent predation and saw marks can appear similar during on-scene visual inspection, but quite different under a microscope.

"Anthropologists get asked to do some bizarre things. Because a lot of our studies include comparative anatomy, we get the 'Was it animal or human?' questions. But we also get the slightly more left-field kind, like 'Was it animals or humans *doing* the gnawing?' "

Facial Reconstruction

Perhaps the most dramatic facet of forensic anthropology is the restoration, often from a bare skull, of a face that is recognizable by friends or acquaintances of the victim. There have always been two methods of attempting these reconstructions. One involves a complex process of layering the skull with clay. The other relies on the combined efforts of a talented composite artist who can recreate faces from the description of witnesses and an anthropologist who can visualize and accurately describe the face from the skull. A third technique—computer-assisted modeling—has taken the anthropologist's skills into the computer age.

Computer modeling programs, which work on the same principles as the other methods, can perform two important forensic tasks. First, they can age or regress a face based on well-established growth patterns. Ears, for example, get longer as we age; so, given the photo of a seven-year-old child kidnapped many years ago, the anthropologist can adjust the image to "age" the subject. By adjusting each feature individually, the anthropologist soon has the basic outline of the older face and begins the delicate process of smoothing the lines, blurring the edges, and eventually creating a realistic face. Given an older face, the reverse process is also possible. This is especially valuable in work with amnesiacs or in cases where a person may have lost contact with his family for many years before dying.

The family's visual memory of the person would be of the younger individual.

In traditional skull molding, the anthropologist begins by identifying the age and sex of the individual—both of which can be determined from just the skull itself if the skeleton isn't available. From measurements of the skull's features, a judgment of racial or ethnic type is attempted and key muscular attachment points are determined. With these points and a good idea of facial type, anthropologists consult a table of average tissue thickness for those points. Pieces of plastic, which look a lot like an elongated version of the eraser at the top of a pencil, are cut to the appropriate thickness and applied directly to the appropriate points on the skull.

Muscle analogs, clay shaped and sized to fit the skull, are laid in place next. Some portions of the face, such as ears and tip of the nose, are almost entirely cartilage, so these are the areas that present particular difficulty. It's impossible to tell if someone's nose tipped up, or if they had tiny, delicate ears or dinner plates sticking off the sides of their face. The area around the eyes is likewise lacking in bony structures, so this feature— the most expressive and individual of a real face—has to be estimated somewhat. Judging the depth of the eye itself and knowing the spacing pupil to pupil—both possible from the eye sockets themselves—gives most reconstructions a more lifelike appearance.

Reconstructions done in clay can be remarkably accurate depictions of the individual, as can the drawings done by a talented team of artist and anthropologist, or by an artistic anthropologist. But being able to reproduce these activities on a computer screen offers some advantages, too.

"We've always said that we have to guesstimate some features," says criminalist Wally Schier. "And the computer modeling allows us to make a dozen faces with all the variations possible for that skull. It also lets us play with skin tone or hair color, things we can't know for certain without some

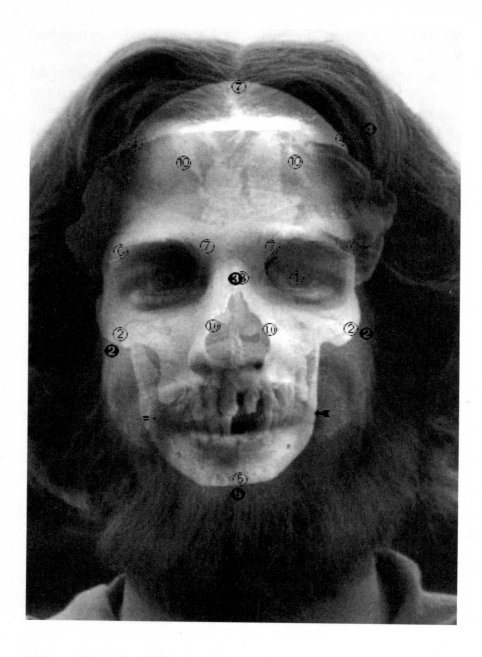

supporting evidence, like associated hair. Instead of being stuck with the model we made today, and having to completely redo it to account for these possible variations, we can do one model, make the adjustments on-screen, and print out all the possibles for investigators."

Modeling has other advantages, as well—distinctly forensic advantages.

"In the process of scanning in the three-dimensional model of the skull—the basis of the completely virtual reconstruction, the type where we don't do anything to the skull itself except scan it—we also end up with high-resolution scans of any injuries to the skull. A hammer makes a particular sort of mark, and the model can be very precise in the contours and depth of that weapon mark. Tool marks visible under a scope like an SEM can be recreated digitally, as well. With good modeling of both the wound and the weapon, it's possible to match them in visual reconstructions of the *event*, not just reconstruction of the face."

Digital superimposition is yet another way anthropologists take advantage of a computer's ability to do grunt work in seconds. "If we present a reconstruction and get a hit, a suggested match, from a member of the public or a law enforcement agency elsewhere, there are often photographs available of the missing individual. With our scans of the skull on file, we can then digitize the photographs and attempt to match them feature for feature with the skull in a series of computer overlays. Even if we can't declare an absolute match, we can certainly exclude possibles this way—which may be just as important in the long run."

The Autopsy

Not all autopsies are forensic. Determining cause of death may be a strictly medical procedure, designed only to answer questions of medical curiosity and not to gather evidence for future legal proceedings.

However, violent deaths, suspicious deaths in apparently healthy individuals, and some accidental deaths, fall firmly into the forensic field.

WE'VE COME A LONG WAY

Anne Perry's meticulously researched historical mysteries include all the details of Victorian life: the recipe for rose milk, (used by all fashionable ladies at bedtime), the absolutely authentic way to remove singe marks from linens, and, in *Ashworth Hall*, the socially acceptable way to conduct an autopsy at a weekend house party after someone starts murdering the guests.

- First, store the body in the ice house where "Carcasses of meat hung on hooks . . . offal sat in trays, and several strings of sausage looped over other hooks."

- Then, ask the victim's son, who just happens to be a medical student, and not even a qualified physician, to perform the cutting.

- Stuff your corpse in a laundry basket to ensure easy handling between the ice house and the laundry room where the autopsy can be performed at midnight when the discovery of a corpse on the ironing board won't excite too much angst in the servant population.

- If the well-stocked Victorian home doesn't include any medical gear, cook can be counted on to have a well-stroped filleting knife that should serve nicely.

All of which would be hilarious—if it weren't completely likely! An 1823 case was brought to court in London during which a "postmortem examination of the decedent" conducted by a "most able flesher (aka a butcher) kindly obliged her husband by returning the cause of death natural on the spot," the spot being his shop table.

The woman was well-buried before the husband bothered to inform authorities that she was dead.

The obliging flesher never did give testimony or appear as a witness, but the court was happy to take the husband's hearsay evidence as proof.

Modern forensic autopsies have several goals:

- determine the identity of the decedent
- determine the cause of death
- determine the manner of death
- determine the mechanism (or mode) of death
- determine the time of death

The weight of each question varies with the case. Some victims are known, others only suspected. A car crash victim's time of death might be known to the second, but other estimates for time of death may cover a period of weeks and still be considered significant.

Cause, mechanism, and manner of death determinations can, together, suggest how a crime or accident may have occurred.

Cause of death is the direct agent that leads to a death. A bullet, poison, and electrocution are all causes of death.

Mechanism or mode of death is what happened to the body as a result of its run-in with the cause of death. If a bullet is swallowed, it may cause no harm at all. If a bullet tears a gaping hole in the aorta, death is almost inevitable. Bleeding into the chest cavity (with its attendant loss of blood pressure and other factors) would be the mode of death.

The manner of death can be any of these:

Accident: A man cleaning his gun accidentally shoots himself in the chest and dies when the bullet opens his aorta, causing enough blood loss into the chest to compress the heart and prevent it from beating.

Suicide: A woman presses a gun just right of her breastbone and fires, dying when the bullet opens her aorta, causing enough blood loss into the chest to compress the heart and prevent it from beating.

Murder: A woman presses a gun to the chest of a man and fires. The aorta is severed, the man bleeds out into his chest, causing sufficient pressure for the heart to stop beating.

Natural: A man is eating breakfast when an inherited weakness in the aorta gives way, causing enough blood loss into the chest to compress the heart and prevent it from beating.

In forensic autopsies, policies and procedures guide the medical examiner and forensic scientists through a process that should allow them to gather the most useful information. However, they still may not be able to answer all the questions given them. Ideally, evidence gathered from the body, from the scene, and from the investigation of the circumstances surrounding the event will provide a narrative of what happened.

WANTED—ASSOCIATE CHIEF MEDICAL EXAMINER

The North Carolina Office of the Chief Medical Examiner is seeking applicants for the position of Associate Chief Medical Examiner.

Qualifications include a Doctor of Medicine degree from an accredited school of medicine; eligibility for a license to practice medicine in North Carolina; completion of residency in hospital pathology and in forensic medicine; and Board certification in Anatomical and Forensic Pathology.

Responsibilities include assisting the Chief Medical Examiner; supervising staff of Pathology Branch (autopsy, histology, and photography services); performing approximately 250 autopsies per

year; reviewing approximately 1,450 filed medical examiner cases per year; and providing instruction in Forensic Pathology.
Salary: $97,704–$170,649/annual.

The medical examiner, pathologists, and other forensic investigators aren't responsible for interpreting the evidence and creating an entire narrative. Their job is to find the evidence, preserve it, and report it accurately; hopefully they'll be able to determine manner of death in the process. Some examination of the body will have already taken place at the scene and those reports will come to the medical examiner. In most jurisdictions, the medical examiner's office attends the crime scene, so the circumstances are fresh at the time of autopsy. Medical histories or dental records may become available if the subject is known. (If the person was alive at the scene and received any sort of resuscitation, was injected, had IV or thoracic lines run, was intubated, or handled by bare-handed medical personnel, that information is vital in excluding later effects on the body as part of the crime.) Exemplars from the scene should be available for comparison to the exhibits that will be generated during the postmortem examination. All this material, as well as whatever parts of the narrative law enforcement investigators have discovered, comes together over the autopsy.

The classic law school enigmas and questions seem perfectly at home in the medical examiner's office. "A man is pushed from the top of a hundred-story building by one party. On the way down, he has a heart attack likely to be fatal. On passing an eightieth floor window, he's shot by a third party and dies before impact. What are the cause, mode, and manner of death?"

To answer equally twisted medico-legal questions, examiners follow a standard protocol in approaching the body:

- visual inspection
- injury inspection
- internal examination

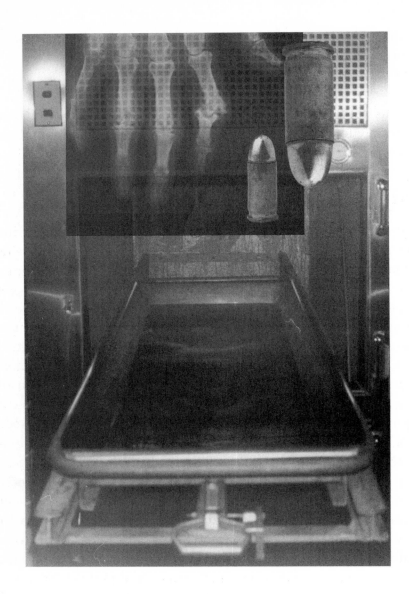

- presentation to other experts
- reconciliation of all exhibits
- presentation of findings

The depth or breadth of evidence believed viable, useful, or appropriate for collection varies dramatically from case to case, but probably includes some combination of the following procedures.

Visual Inspection

Even before the medical examiner starts a visual inspection, some preparations may have been made. Some offices draw blood on receipt of the body (simply because, at this point, it may still be somewhat fluid); other offices wait to collect samples at the PM; they may also X-ray, weigh, and measure the body before beginning the postmortem exam. X rays reveal not only broken bones but also foreign objects that are potentially hazardous to the examiner. Many times during the traditional wars of the twentieth century, unexploded ordnance of various types was discovered in both living and dead bodies—often with fatal results to the doctor or examiner. In modern times, that is less of an issue, though not totally impossible.

Other potentially hazardous situations do exist, however. Sharp objects not obvious on visual examination—such as the broken tip of a knife left in a wound—can slice through gloves, exposing the examiner to biological contamination. And if bullet fragments have shattered throughout a body, X rays before and after examination help ensure that all possible exhibits have been recovered.

The visual examination for a forensic autopsy can begin with a clothed or nude body, depending on the circumstances at the time of death. In either case, the entire body is photographed, with special attention paid to any unusual details. Rips, tears, or trace on clothing are documented. Trace is collected for presentation to appropriate examiners. Clothing is removed and each piece is air-dried and packaged separately for processing.

Accurate descriptions of each item of clothing are part of the autopsy record and should include any information on the tags, including size and brand. Blood or other stains on the clothing will be handled by the appropriate examiners. Blood and other stains on the body itself will likely be swabbed on the chance that it *isn't* the victim's.

Visual inspection continues with the following:

- Head hair is combed for trace, foreign hairs, and other evidence. Further samples from all over the head may be plucked to ensure that hair examiners have known victim exemplars.
- Depending on circumstances, a standard rape kit may be included in this phase of the autopsy.
- The interior of body openings—ears, vagina, etc.—should be examined for foreign objects or trace evidence.
- Hands, which may be bagged at the scene to prevent loss of possible trace evidence, are examined for damaged nails, and nail scrapings are collected. GSR collection is undertaken if indicated and not already done.
- Fingerprint and footprint cards are completed.
- If not already done, entomological samples may be collected.
- Notations of visual injuries are made and the injuries themselves are photographed.
- The overall condition of the person (usually expressed as "healthy" or "undernourished" or something similar) is noted, along with hair and eye color.

The visual limitation of human eyes can be overcome with the assistance of specialized lighting or photography. In cases of suspected abuse, UV photography often reveals the remains of old bruises, or new bruises not yet visible on the surface.

Lasers, ALS, and simple oblique lighting turn up trace such as hairs, pollen, blood, saliva, semen—even fingerprints. All are documented and, where possible, collected and submitted for expert analysis.

FINGERPRINTS ON BODIES

The ability to find and collect fingerprints on bodies—the living as well as the dead—links perpetrators and victims as can few other pieces of evidence. Until recently, it was an investigator's dream, but little more. The first successful recoveries were obtained with either an iodine method or a direct transfer method. The transfer was to photographic-type papers pressed against skin; this produced a reversed image. Post-processing was required to correct the image and flip the white/black coloration to normal.

Skin produces sweat, salt, and oil. Most fingerprints are composed of sweat, salt, and oil. How difficult is it to separate the victim's sweat, salt, and oils from the suspect's? An investigator once described it as trying "to lift ripples from a puddle."

Skin itself isn't the best surface under any circumstances. It stretches, which means that prints can be so distorted it's impossible to make any match. If the victim doesn't die (obviously no investigator or examiner *prefers* a dead victim to a perfect print), skin (a living organ that constantly renews itself) might shed the layer where investigators are hoping to find the print. It's not at all unusual for areas of contact between suspects and victims to include injuries, and injuries are usually treated by a variety of people (who often leave prints of their own, or smudge existing prints) or subjected to cleansing prior to necessary medical treatment. All of these things destroy prints.

In the early 1990s, overcoming those problems became a group effort on the part of the FBI, the University of Tennessee's Anthropology Department, the University of Tennessee Hospital, and the Knoxville Police Department. Using real cadavers and natural prints (other studies have used embalmed bodies and baby oil to simulate reality), they tested dozens of possible methods and materials. Eventually, they declared cyano-fuming with secondary enhancement to be the most consistently successful method.

Clearly, fuming an entire body wasn't always practical or desirable, and the old problem with the cyano techniques—the fumes can send even experienced examiners out of the room with severe headaches—prevented the sort of predictable results necessary for forensic work. A police specialist from Knoxville eventually created a portable fuming unit that allowed criminalists to fume specific areas of the body with a consistent amount of cyano fume for very specific periods of time. (If you recall, too much fuming results in a mucky white blob.)

In cyano-prints on bodies, researchers also discovered that simpler was often better and that regular black magnetic powder could produce results every bit as useful as the more exotic powders. This makes it possible to respond to more situations with only the usual fingerprint kit contents.

During the field testing of the technique, it became obvious that the resulting prints were stable. The body could be removed from the scene, and the powdering and lifting could be done later. Fieldwork also showed that, given any sort of break in environmental conditions, older prints might also be lifted.

At long last, the black pit of fingerprints—the human body itself—had been illuminated.

Body mark documentation (essential information if subject identity is yet to be determined) includes photographing and sketching any scars, tattoos, or moles. An unknown's tattoos, in particular, should be circulated in case it's possible to identify either the artist or the meaning of the design. Military units, fraternities, jail blocks, teams, and other groups might share common tattoo designs. Truly unique work (as opposed to the sort plucked at random off a parlor wall) may be attributable by theme or style to an artist, narrowing the geographical location and perhaps even resulting in a direct identification.

Scars from medical procedures—useful in identification cases—often survive abrasion and even burns as the scar tissue penetrates *all* layers of the skin, and not just the upper layer, or epidermis.

Birthmarks and other distinguishing marks, frequently noted

on passport applications, and certainly known to family and closer friends, can be distributed and recognized even when portrait photos elicit no response. Dead people rarely look all that much like themselves.

What bodies frequently *do* look like is the surface they lie on at the time of death, when lividity (the discoloration of the skin as blood settles to the lowest points in the body) can be read most easily. The impression of rocks and bottle caps—even the print of a tire jack—can record the place where this individual lay for at least some portion of the time since the moment of death. In Australia, a body dumped along a rural road led investigators directly to the murderer; the body's lividity revealed the distinct imprint of a license plate found in the trunk of the vehicle that had been used to transport the corpse. The killer had taken off the plate so that anyone seeing the car in the area wouldn't have a plate number to repeat to police. The body lay on the plate for the entire trip to the countryside. This exact scenario, though enacted in the setting of urban Boston, was recently incorporated into a story line for *Crossing Jordan*.

Injury Inspection

An important part of any visual survey is injury inspection. Injuries are broken down into two basic types: penetrating and non-penetrating.

Bullets, knives, and bombs all cause penetrating wounds, some featuring both entrance and exit openings. The full extent of these wounds can't be described from an external visual survey, but there is still plenty of important information available at this stage.

Concerning bullet wounds, the first thing noted should be the presence or absence of gunpowder marks around the wounds. Another moment of examination can reveal much more:

- A bullet wound from a weapon held some distance from the victim results in a bullet hole, but leaves no other visible surface information.
- A wound from a weapon held directly on the skin produces the bullet hole and a "rim burn" directly around the opening. This is a result of the heat and flare of the shot.
- A weapon held just a few inches back from the skin produces a noticeable pattern of soot around the wound, but no burn. Hair is often scorched while skin remains unburnt, simply because the outer layers of hair are extended those few inches from the scalp.
- Depending on the weapon, a weapon held anywhere from another few inches to a few feet back will result in "stippling"—the almost tattoo-like markings of the particulates carried by the cloud of gas actually embedding in the skin. Beyond that distance, the particulates disperse in the air between shooter and victim.

Because of the elasticity of the skin, the bullet wound itself can change shape, so estimates of caliber aren't usually possible. But a few other bits of information can be pulled from the surface appearance of bullet wounds. In the case of the through-and-through wound, entry wounds are generally smaller and neater than exit wounds. A contact-type wound directly over bone (as opposed to soft tissue like that of the abdomen) may show the star pattern—a circular wound punctuated by radiating breaks. So-called execution-style wounds to the back of the skull often display this pattern.

REALISM IN WRITING?

Reading audiences, exposed to some meticulously researched crime fiction that is often written by experienced lawyers, police officers, and medical personnel, have come to recognize "fudging." Few would stand for a gunshot wound description like this one by Marcia Cloud in *The Cabal*:

Her eyes fixed on the victim, the tiny hole, the rim burn a red bull's-eye surrounding it. The entire side of the boy's head, covered in distinctive gunpowder tattooing, its hair burnt away, left her shivering.

It should have left her confused!

If the gun was far enough from the skull to leave stippling over "the entire side of the boy's head," it was clearly too far away for any encircling burn from a gun muzzle pressed to his head.

Cloud, now writing under another name, freely admits, "I didn't really understand any of that. I more or less copied bits from several court transcripts. Not a mistake I'd make today."

But this description from Michael Connelly's *Trunk Music* draws approval from both savvy readers and medical examiners:

In the mess that was the back of the man's head, Bosch could see two distinct jagged-edged penetrations to the lower rear skull—the occipital protuberance—the scientific name popping easily into his mind. Too many autopsies, he thought. The hair close to the wounds was charred by the gases that explode out of the barrel of a gun. The scalp showed stippling from gunpowder. Point blank shots. No exit wounds that he could see. Probably twenty-two's, he guessed. They bounce around inside like marbles dropped into an empty jelly jar.

Keeping in mind that a "point blank shot" isn't a contact wound, merely a close one, Connelly got it just right.

Knife wounds and other similar penetrating wounds are three-dimensional and can often be cast to reveal the shape and size of the weapon. But experienced examiners can frequently make a preliminary description of a knife, awl, or other object based on the surface of the wound. A knife hilt or object handle pressed forcefully into the surface of the skin leaves marks. Most knife assaults use common kitchen ware, which has just the one sharp edge, and a single-edged blade makes a penetration wound different from a double-edged blade.

Depending on the item used, blunt-force trauma may leave

characteristic marks. Hand and finger impressions around the neck or face can be measured for finger span and size. Bite marks certainly require photographing but, depending on the situation, may also suggest special castings to capture a 3-D representation.

Many environmental and traumatic factors leave visible marks on the skin's surface. Burns and frostbite both leave reddened areas across the skin. Deliberate burns from cigarettes can look like bullet wounds on first glance, but look considerably different up close. Electrical burns usually occur in pairs— one from the point of contact, the other at the grounding point where the current passed out of the body. It's not unusual for both hands to be burned. (Electrical injuries cause some other odd happenings inside the body, as well. The shock can cause muscle spasms so severe that bones are broken. The chemical response can lead to kidney failure and epileptic-like seizures after the fact, separating the cause and method of death by days or even weeks.)

Visible marks on the skin can be especially important in

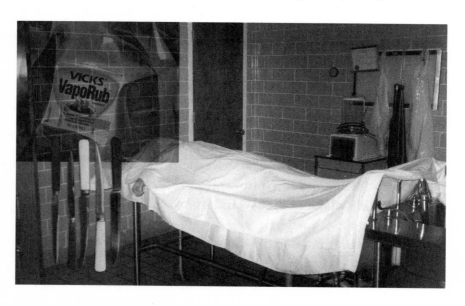

determining manner of death in strangulation cases. For example, in hangings in which the body dangles from an overhead rope, the rope mark around the neck is usually characterized by an upward slant. On the other hand, garroting (a faked suicide setup) would leave a horizontal line.

Other injuries to the body surface point to new avenues of investigation. Numerous needle marks on the arms or legs or between the toes, for example, certainly suggest drug use. This should be cross-referenced with medical records (if they're available) to provide a time frame, and flagged for forwarding to the toxicologists who will handle the laboratory aspects of the autopsy. Singular needle marks might point an investigation in an entirely different direction.

When every square inch of the body surface has been examined and documented, and every scratch, bruise, bump, mark, and wound has been sketched and photographed, the body is carefully washed (this waste water may be examined, as well) and briefly reviewed to ensure that nothing was missed under blood or other debris. With clues from the external exam fresh in mind, the internal survey can begin.

Internal Examination

Step one of almost every dissection begins with the "Y" incision over the torso. One cut—an arc swooping down across the chest from shoulder to shoulder, down to the base of the breast bone—forms the upper extensions of the Y; the lower extension is the longer cut—from the base of the breast bone to the pubic bone. If there is obvious injury to the extremities, or any possible medical condition that would show evidence in the limbs, these will be examined later.

With the torso laid open, organs are removed one at a time, starting at the top and working down the body. Rib cutters and spreaders open the chest itself to examination. If blood hasn't been obtained from the victim at this point, heart blood will be drawn now. A gross examination of the lungs reveals the presence or absence of water, clots, and corrosive materials. Any fluids in the lungs will be sampled and sent to the lab. After physical examination, the heart and lungs will be weighed, then sectioned for microscopy if indicated. The throat, esophagus, trachea, and upper spine can also be examined at this time. The mouth and nose are part of the external exam, as are dental surveys.

Stomach contents, which can help determine the time of death and tie victims to particular places, also go to the lab, though most examiners can determine something about contents on visual survey. The presence of any drugs, or the suspicion that drugs might be found in the stomach or intestines, requires that toxicological studies be run on the contents as well as the blood. Differences in concentration throughout the body help pinpoint when the drugs were taken and if ingestion was habitual. Also, if the drugs haven't had time to disperse, this can help fix time of death.

Each organ, such as the liver, spleen, and pancreas, can be sectioned for histology and pathology study. Gross injuries or bruising may be indicative of blunt force trauma. Penetration wounds can be measured and checked for angle of entry and exit. When all organs (including the adrenal glands and kidneys) are removed, the next section of spine can also be examined, if appropriate. Again, each organ is weighed and examined individually. A liver sample is routinely taken. Like the stomach contents, contents of the intestines can help establish time of death. Also, the presence of any undissolved drugs in the intestines might be significant.

Last in the torso is the pelvic region. In both men and women, fluid from the bladder is recovered for drug analysis. The presence of blood can point to blunt force trauma over the kidneys, if such has not already been discovered. In women, the

reproductive organs are examined for signs of pregnancy, sexual interference, and, in identity cases, evidence of previous pregnancies, births, or abortions, all of which help narrow the field. If a pregnancy is discovered, the fetus will be examined, as well, but more on that in a moment. Should semen turn up in the examination of the genitalia, it is of course documented.

The path of a knife or other object wound through bone or cartilage may contain the equivalent of tool marks. If found, the section of hard tissue should be completely excised for photography, possible molding, and examination under high-power magnification.

When the torso examination is complete, any sections that were removed are returned to the cavity. How they are returned depends on the policy of the regional authority. In some jurisdictions, the organs are bagged first. If an individual organ shows some unusual property or characteristic, it may be preserved whole and retained by the examiner or lab for further study.

The second part of the internal examination focuses on the head. First to be examined are any peculiarities on the eyes or eyelids. Samples of the vitreous (the fluid of the eye) can be aspirated and forwarded to the lab. Hemorrhages to the eyes or lids are noted, as they can indicate interruption in circulation to the head. Hanging or strangulation might account for them, but it's also turned up in cases where neither of those were indicated. A man in Maui once displayed such distinct red spots after a prolonged bout of sneezing, while a Connecticut woman exhibited the same after a particularly long labor!

Once the face and eyes have been thoroughly described, a single incision—arching from ear to ear over the top of the head—is made. The face is folded forward. This exposes the skull, which is examined for breaks, bruising beneath the skin (often hard to see due to head hair), and any sites where the skull itself has been punctured. The skull is opened with a saw, and a visual inspection of the brain is begun. Clotted or unclotted blood inside the skull should be noted and matched, if possible, with injury indicators on the exterior of the skull. The brain is then removed, weighed, inspected more closely, and sectioned for microscopy.

Any marks to the extremities, wounds, and entry or exit points are investigated next. Some details are more likely to be visible on the extremities than on the head or torso. This includes defensive wounds, which are usually found on the hands and forearms, but are also occasionally found on the legs if victims kicked at their attackers. Damage from freezing may be evident in the fingertips or toes. The "Dorian Gray Effect"— the incredibly rapid breakdown of a previously frozen body— can be most obvious in the extremities.

COMMON SENSE HELPS

An early episode of the *The Saint* (the Roger Moore version) created an unusual situation that would leave any medical examiner baffled. The fictional doctor, removing a slug from the shoulder of a Nazi hunter, pulled it from the entry wound—rounded-tip first.

Leaves one to wonder how it managed to turn around in midair and enter backward!

Medical examiners confront unusual situations regularly. Interpreting the conditions of bodies that have been recovered from water, bodies that have been burned, and bodies that have been separated into several pieces requires special knowledge, but a few circumstances require special handling from a legal standpoint, as well, such as the following:

- fetal deaths
- exhumations
- partial bodies/amputations

Fetal Death

There are three scenarios, from a legal perspective, for a fetal death:

1. *A fetus of less than 500g is "miscarried," expelled from the mother's body through natural or traumatic means.* (From country to country, and state to state, there is variation in the exact weight determination.) Generally, no forensic autopsy is conducted and no death certificate is issued. A possible exception might exist if the fetus were considered "motive evidence" in an ongoing case. Blood or DNA typing to determine paternity may be valuable to the investigation.

2. *A fetus of more than 500g is born dead—a "stillbirth."* In this case, in the absence of violence or suspicion, a special fetal death certificate is issued with cause of death given as "non-viability of fetus." Forensically, there is no method or manner of death, though a general autopsy may be conducted to answer medical questions.

3. *A fetus of more than 500g is born alive (for any period of time) and then dies.* This situation requires that a standard death certificate be issued and either a forensic or general autopsy may be performed, based on the scenario as interpreted by criminal investigators and medical staff.

As fetal death may not be an isolated event, examiners and investigators may be confronted with a combination of deaths that includes the mother's. If a mother and fetus (over 500g) die, then two standard death certificates are required in most jurisdictions. And, if the deaths are held to be violent or suspicious, forensic autopsies of both bodies are required. The methodology used in a fetal forensic autopsy mirrors that of an adult autopsy.

Exhumations

If, for whatever reason, a body already buried is required for forensic study, then it must be exhumed. Exhumation requires that investigators present the courts with good reasons for disinterring a body, and an exhumation order must be obtained from the court before recovering a body.

Death processes in buried and/or embalmed bodies don't follow the natural course; this presents serious challenges for examiners. During embalming, blood is drained from the body and replaced with a formaldehyde compound that often contains pink dyes to improve the appearance of the body. This mixture preserves the body for a time, but drastically alters its chemistry.

Depending on the length of burial time, remains may have completely skeletized and examiners may require the assistance of an anthropologist to make meaningful observations.

Partial Bodies/Amputations

The object of any forensic autopsy is to identify the individual, as well as the time, cause, method, and manner of death. It's not easy to achieve those goals under the best circumstances. But given only a partial body, or an amputated limb, the deck is stacked against the examiner.

"We once spent seven months trying to assemble the body of a single person," recalls medical examiner Jean LeClerq. "We estimate she was hacked into forty-two or more pieces in total. We never found the head, hands, or feet at all, and suspect they were burned or something. We know the murderer was counting on predation to scatter the pieces beyond our finding. Damn near did, too. We recovered something less than fifty percent of the body, even after intense searching."

The badly decomposed pieces were subjected to intense examination individually, with ID coming from a combination of medical screws in a long bone, a question-mark–shaped scar on the hip ("from backing into a nail when the victim was a child"), and a surgical scar just below where an elbow should have been.

Presentation to Other Experts

Even TV's Quincy consulted other experts and, in the real world, medical examiners aren't expected to recognize every type of trauma ever inflicted on the human body, every agent infecting it, and every poison ever pumped into it.

Histology, toxicology, serology, and other specialty departments regularly receive, test, and produce reports on their findings of samples taken during autopsies. Each report is copied to the examiner and any law enforcement or medical officers with a need to know.

{ "Like I said before, arsenic's a natural element. It doesn't disintegrate, doesn't flush out of the body, so there's buildup. In excess amounts, the first place it's usually laid down is in the hair and skin because of the rapid cellular turnover. To really see how long the poisoning had been going on, I'd have to section Jupiter's bones. Bone growth is like tree rings. If there's poison in the bones, you know it's been going on a long time." }

Jupiter's Bones, Faye Kellerman

In unusual deaths, experts in a number of non-medical fields may need to be consulted. Scuba-related deaths result in bodily changes not normally seen at autopsy, and so the consultation of experts is required. Chemists and pharmacologists assist in the identification of unknown drugs or corrosives. Animal bites or stings require input from outside experts to determine not only *what* created a particular mark, but what possible evidence might be preserved *from* such marks. (For example, venoms, in particular, collected in time, contain considerable information.) Reports of consultations and the results of those investigations all become part of the autopsy record.

WANTED—FORENSIC CHEMIST

The Department of Justice, Drug Enforcement Administration, is seeking applicants to fill seven positions as Forensic Chemist at various laboratories.

Qualifications include a bachelor's degree in a physical or life science, or engineering that includes 30 semester hours in chemistry, supplemented by course work in mathematics through differential and integral calculus, and a minimum of six semester hours of physics. Should have a combination of education and experience, coursework equivalent to a major as shown above, including a minimum of 30 semester hours in chemistry, supplemented by mathematics through differential and integral calculus, and at least six semester hours of physics and appropriate experience or additional education.

Responsibilities include performing chemical and physical tests and instrumental analyses to detect the presence of a controlled substance and determining its concentration in a drug sample; establishing the identity and concentration of accompanying adulterants and diluents whenever possible; advising and assisting in the performance of certain enforcement activities; operating analytical instrumentation such as IR, UV, and fluorescence spectrophotometer, gas chromatograph, and GC/MS; revising and developing procedures as necessary to accomplish the analyses of more complex drug mixtures or trace quantities of a particular substance; writing laboratory reports describing all tests performed, calculations, and

conclusions; testifying as an expert witness; and assisting prosecuting attorneys in preparation for technical aspects of cases.
  Salary: $28,535–$54,987/annual.

Reconciliation of All Exhibits

Prior to the completion of the autopsy, all samples, reports, and exhibits are accounted for and double-checked for accurate referencing. No one wants to be in the position of being asked to produce the original sample and discovering that it's at some other lab, or lost. If a bullet went for ballistics testing, then a chain-of-custody note must be included in the reconciliation of exhibits. If notes were made of samples sent to toxicology and no results have been returned, this is the time to track those reports down.

Any unanswered or unusual findings may require further investigation, or may be deemed unanswerable at this time.

Presentation of Findings

In a forensic autopsy, the final presentation of findings is the autopsy report itself. This report describes all the findings, the steps taken to answer any questions arising from the findings, and, finally, the opinion of the examiner on cause, method, and manner of death. In some cases, the manner of death may be left as "Pending," but the cause of death should be known. From this report, the information for the death certificate is drawn and this document is prepared.

The last step is to release the body to the appropriate authority, or to retain it as evidence if indicated.

FOUR

WORKING THE SCENE: DIFFERENT STAGES

Whether it's a bank vault or a body, every crime scene gets treated to the same procedures. These are the basic steps to be followed:

- Secure the scene.
- Ensure the safety of all persons, those already on the scene as well as anticipated arrivals.
- Administer medical assistance as required.
- Secure witnesses, victims, and suspects, and do so separately.
- Set up a point person and initiate activity logs.
- Secure all evidence.

That sequence never changes, but it expands and adapts to accommodate vastly different crime scenes and investigative techniques. The following sections illustrate the special needs of two contemporary crimes—one that is physically massive (bombs and explosives), and one that is physically minuscule (computer crime).

Bombs and Explosives

Events like both World Trade Center bombings and the assault on the Alfred P. Murrah Building in Oklahoma City graphically displayed the devastating potential of bombs. Still, huge events like these have failed to alert the United States to the *daily* instances when emergency and investigative personnel across the country respond to bomb and explosive threats. According to a National Institute of Justice report, some 38,362 explosives incidents were reported between 1988 and 1997. That's an average of 4,262 per year, nearly twelve incidents every day, one incident every two hours.

Those are staggering numbers when you realize that they are drawn from a variety of databases and reflect *only* those events *reported* to the national agencies. Actual events are suspected to be as much as 30 percent higher because many smaller jurisdictions don't report hoaxes, false call-outs, and other "nonevents," even though these incidents waste resources and make specialists unavailable for real events.

Though film dramatizes bombs as terribly sophisticated devices—*Die Hard III*, *Live Wire*, and *Blown Away* being typical examples—that's not borne out by real-world statistics. The bombs and explosives of choice in America are homemade implements: Molotov cocktails, and pipe bombs. There are almost never any exotic chemicals or accelerants involved. Gunpowder, nails, common gasoline, and other items found legally all over the country make up most explosive and incendiary devices. Fertilizer–gasoline combinations, still a very basic sort of technology, are rare. Equally rare are devices composed of actual commercial blasting material, though every state in the Union reports thefts of these components every year.

While we've come to associate explosives with political statements, and news headlines both international and domestic regularly feature one group or another claiming responsibility for such events, that's not been the case in the United States. The majority of motives in the use of these handmade, homegrown events appear to be personal revenge. The leading targets are equally personal locations: homes, cars, and mailboxes.

One thing appears true of all bombings, regardless of target,

motive, or size: they're meant to do damage to people and property. No one sends a bomb to *scare* anyone. The general theory among those studying bomb scenes is that every bombing utilizes the maximum technology and amount of explosive available to the bomb maker; self-restraint doesn't appear to be a characteristic of bombers. That desire to cause as much damage as possible to the greatest number of people or property makes bomb scenes some of the most difficult for investigators.

Of primary concern is the trend to set bombs and incendiary devices in a series or in stages. The first explosion brings emergency personnel racing to the scene just in time for secondary explosions to engulf them as well. All of which means that bomb scenes must be worked with even more attention to securing the safety of incoming responders and those already on the scene. In some cases, it means obtaining the assistance of agencies not normally involved in criminal investigations: building inspectors, structural engineers, dog handlers, explosives experts, gas and utility personnel—even biohazard or hazmat teams, depending on the location of the explosion.

"The natural instinct of everyone responding to an injury scene is to get to those people, and it's heart-wrenching to stand outside a piece of tape listening to people crying out for help and having to tell your people to stand fast while others create a secure path to them," says DCI Declan O'Neill, who clearly recalls the first day he had to do just that. "In Europe, most bombings *are* political, and emergency personnel are seen as 'the enemy' or 'agents of the enemy.' Once you see the guy next to you blown up while lugging in his medical kit, you realize that, tough as it is, all those other people have to do their jobs, too. Anything else just ups the body count. And now we're starting to see the same strategies in personal attacks as in political."

Working a scene with the specter of further explosions hovering can't make the meticulous tasks required of crime scene investigators any easier. But even without that added stress, these scenes, which can cover acres in the worst-case scenario, are inherently tough. Scene of Crime Officer Tony Sinclair has picked apart more bomb scenes than he cares to remember. "But they're all the same. A lot of dust, tiny particulates, lots of chemistry to run, lots of sifting, lots of wondering if this little bit of wire means anything or if it's the inside of someone's clock radio. On good days, you don't have to bag pieces of people, too."

Large scenes, often outside, make environmental factors important considerations. A rainstorm or strong winds drastically alter the crime scene landscape. Trying to ferret out the important evidence before scene degradation advances—but without destroying more information—requires experienced eyes, and a lot of them if the scene is complicated or large. In large scenes involving unrecovered victims, investigators are exposed to possible blood-based pathogens present in victims, to the biohazards of decomposing bodies, and to the scavengers attracted to human remains.

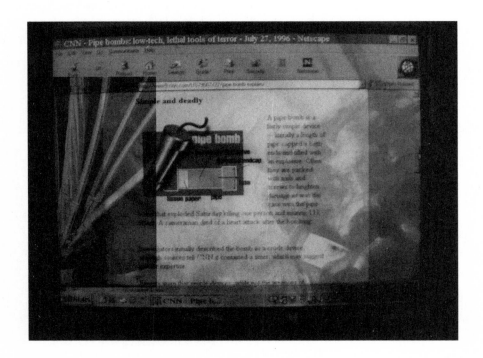

Equipment for explosive and incendiary scenes is highly specialized. Vapor detectors, climbing gear, GPS, and trace explosives collectors/detectors are required, in addition to the usual scene kits. Few other scenes require you to bring such heavy equipment to work.

A DIFFERENT SORT OF CRIMINAL?

"The old saw about the crooks returning to the scene of the crime probably has never had more going for it than at explosion sites," notes Tony Sinclair. "I'm not sure who first pegged it, but about 1992 it was suggested that an awful lot of convicted bombers hung about the scene afterward, sometimes trying to chum up with the volunteers or bring food to the emergency crews.

"Arsonists and bombers seem to be hang-abouts–even more than your average criminal."

Which explains why one of the additional personnel often asked to attend bomb and arson scenes is a videographer or photographer. He or she doesn't record the scene, but the people on and around the scene.

"In several key cases, it's been those images that have led us to the person or persons ultimately convicted."

The initial walk-through at a bombing or incendiary event has specific goals:

- establishing a "grid," or three-dimensional search pattern, to find the smallest evidence
- documenting secondary evidence of blast patterns, including (but not limited to) deformed signs, melt patterns, structural damage
- preparing a preliminary assessment of possible bomb delivery methods
- establishing the number and location of blasts and determining if all blasts resulted from devices or included secondary blast origins, such as gas lines or stored materials

- ensuring that all emergency medical personnel are aware of the need to perform full-body X rays on all victims (dead or living)

The scene will require all the evidence collection practiced at non-explosives scenes (fingerprints, documents, blood/hair/fiber), but investigators must also attempt to obtain and preserve these:

- possible bomb components
- items associated with bomb making or placement (placement plans, wires, molding plasticine, batteries, vehicle parts in the case of car bombs)
- swabs of all surfaces for chemical analysis
- exemplars of local materials
- evidence of victim placement prior to the explosion

HOW NOT TO TEST RESIDUE

C.S.I. normally takes some pains to get forensic details right, but in the episode "Boom," the lead investigator did something that left Danny Thomas, a former member of the Clarke County Fire Department, staring at his TV screen. Gil Grissom deliberately burned off a suspect residue collected from a body, and burned it off directly *over* copious quantities of the gunpowder residue **still** covering the body!

"No, definitely not. Nobody plays around like that."

When the scene has been processed, the evidence evaluated, and the reports filed, bomb and arson investigators still have another task to do. They must submit information on the dozens of exhibits collected to national databases that might be able to suggest similar methodologies or similar preferences in components and materials.

The major reporting authorities include the FBI Bomb Data Center and the ATF Arson and Explosives National Repository.

THE INDISCRIMINATE KILLER

Bombers and arsonists, like spree shooters and those who tamper with food and drug products, are indiscriminate killers. Even if their target is specific, they're willing to kill unknown numbers of complete strangers to reach it. More often, however, the target is nebulous.

The Unabomber took a political stance and sent bombs to any individual he felt violated his beliefs. He did not need to personally know the recipient, and he wasn't particularly concerned if someone other than the target were injured.

Timothy McVeigh brought down a federal building and seemed unable to differentiate between the government itself and the children of government workers who died in the same incident.

For the indiscriminate killer, targets may be completely symbolic, making it difficult to link crime and perpetrators as, in many cases, the suspect's symbolism seems indecipherable to the rest of us.

Computer Crime

Computer crime scenes are complicated by the transient nature of electronic evidence, the speed of information transit, the possibility of sophisticated encryption, and the anonymity available to users.

For the crime scene investigator, simply recognizing what material may provide evidence and figuring out how to preserve it requires constantly changing policies and protocols. "Computer crime" incorporates an incredibly wide range of activities. Officers in any unit might be confronted by the need to collect electronic evidence—everything from images inadvertently captured with digital cameras to the list of numbers in a cell phone, or an entire computer network that likely includes dozens of Internet connections. Simply turning off a machine or letting its battery go dead during transit can lose valuable information; consequently, computer crime scene specialists are quickly becoming part of every response team. In the wrong hands, electronic evidence can prove more fragile than any

dusty fingerprint. Even the right hands may not be able to pull all available data from the growing variety of digital gear.

The pervasive nature of computers at modern crime scenes is illustrated by their unique position as criminal tools, as law enforcement resources, and even as *targets* of crime. Carly Schelly, a scene-of-crime officer, once found herself eyeing a setup for an online auction scam and "really, really scheming to get it out of the evidence room and onto my desk! I was positively green! Primo equipment I might see about a year later, if I was lucky."

BITS AND BYTES IN CYBERSPACE

One of the first non sci-fi novels to integrate forensic computer science into a traditional police procedural was *Angels Flight* by Edgar Award–winner Michael Connelly, creator of Harry Bosch. One of his characters, detective Kiz Rider, speaks plain English in explaining how she's connected the case they're investigating to an online porn ring:

> *They [the forensic comp scientists] did some snooping around after getting into the server. They found a cookie jar*

on the Web site. What that means is that there is a program that captures data about each user who enters the site. It then analyses the data to determine if someone has entered the site who should not have had access. Even if they have the passwords, their entry is still recorded and a data trail called an Internet Protocol address is left behind. It's like fingerprints. The IP, or the cookie, is left on the site you enter. The cookie jar will then analyze the IP address and match it to a list of known users. If there is no match a flag is raised. The site's manager . . . can trace the intruder. Or he can set up a tripwire program that waits for a return visit from the intruder . . . the program will attach a tracer which will provide . . . the intruder's e-mail address. Once you have that you have the intruder cold. You can identify him then.

A sampling of possible sources of electronic evidence and a limited list of the crimes they might involve includes these items:

- Address books, Internet bookmarks, and user files can all contain victim lists or contact lists in any number of monetary scams and can provide evidence of contact between suspect and victim in stalker crimes.
- Digital images and movies can be important evidence in certain pornography and pedophilia cases.
- Phone and fax records have been entered as evidence of conspiracy in numerous crimes.
- Way point coordinates from on-board info systems in cars and Global Positioning Systems feature in some drug-related activities.

A computer CPU and a deskful of diskettes and CDs are obvious sources of potential electronic evidence. But less visible elements include these:

- smart cards and dongles, either of which can contain keys to encryption codes, passwords, authentication information, or access codes to information otherwise unretrievable
- phones, caller ID devices, answering machines, and electronic message systems, which can provide a variety of information including potential voice print material, call lists, and the last number called
- handheld devices like Palm Pilots, which can contain all the usual information about a desktop or laptop unit, but may also include handwriting samples and calendar events
- memory cards/devices, which come in dozens of formats, including some that hide inside cameras but don't always record just images, can secrete voice files or timelines in the date stamping connected to most files
- servers, switching hubs, and LANs, which often issue cookies and other temporary files that, if found on particular machines or devices, can lead to logs on the server which, in turn, connect one machine with the addresses of all the devices it may have contacted

- printers, which also contain user logs in onboard memory, as well as incoming fax information if the computer was set to automatically receive and print signals; like typewriters, printers may contain impression evidence, as well
- storage devices like Jaz cartridges, which don't look like other disks but contain many times as much information; also, other storage media, like tape, which can be overlooked, especially if mixed in with audio tapes
- scanners, which don't normally contain information themselves, but may contain smudges or imperfections that may well appear on the images created and saved as digital information
- multifunction watches, which frequently include calendar, address, and even E-mail information, as some can currently connect to computers

- fax machines that can scan several pages for later sending as a single document, with memory that can be probed for those images, as well as log information
- PCMCIA cards, which are about the size of a credit card (though a little thicker) and fulfill a variety of functions; some are modems, others act as memory devices, spare drives, and fax links—all potentially contain electronic evidence

Additionally, the presence of dozens of other electronic gizmos may indicate particular types of crime. Credit card swipers and programmers, cellular phone cloning equipment, CD/DVD copiers, and GPS units that have been used in the human smuggling trade are commonly found at crime scenes. Many gadgets provide not only electronic information, but also physical evidence, so their investigation may require complex interdisciplinary approaches that don't destroy any potential information.

THE CARDINAL RULE

"If it's on, leave it on. If it's off, leave it off. But photograph everything!"

Capturing images on computer screens may be the only way to secure some evidence until specialists can arrive. After moving to the scene-of-crime's computer division, Carly Shelly was told, "If you think you'll have any difficulty remembering *that*, just let us know and we'll have it carved, burned, or tattooed into your forehead for you. Your choice, of course."

Collecting information at a crime scene containing electronic evidence requires that some extra steps be added to the usual routine:

- Determine what electronic items are active either by listening for sounds, physically checking for warmth associated with activity, or finding activity lights, and document the connections to and from them. A simple schedule program means it is entirely possible for criminal activity to continue without any operator present. Operators at remote sites can destroy data about to be collected on the terminal. Gadgets can be physically attached through cables, telephone connections, or infrared signals.

- Document any signal lights on the face of the equipment, any images on screen, any sounds of activity. Check for activity lights on modems. If photography can't capture on-screen activity, videotape can be used.

- Electromagnetic fields of many types can destroy electronic evidence and can come from such common crime scene equipment as radio transmitters, car seat warmers, audio speakers, the magnets used to recover bomb or metallic trace evidence, and the magnetic wands/brushes used in fingerprint technology. *Never* use magnetic powders around electronic evidence.

- Because passwords and access codes can resemble any combination of letters or numbers, it's imperative that *all* written material be available to the computer crime scene specialist for possible use in accessing hidden, encrypted, or off-site data.

- The electric fields that can disrupt electronic information can be generated by regular plastic evidence bags. Only paper and antistatic plastic can be used for computer gear or storage media.

- Cold, dry conditions generate object static. Excess heat can destroy data. All computer-related materials should be kept at moderate temperatures. Secured storage areas should include antistatic mats or touch spots for further protection.

- Batteries found inside units maintain "housekeeping" functions like preserving date/time information and user logs. Evidence custodians must be made aware that long-term storage can deplete these batteries and result in lost data.

POSSIBLE CATEGORIES OF COMPUTER CRIME

Child Abuse (or Exploitation through Porn)

Computer Intrusion (Hacking)

Electronic Stalking/Harassment

Gambling

Smuggling

Telecommunications Fraud

Fraud

Counterfeiting

Identity Theft

Prostitution

Cellular Cloning

Software Piracy

FIVE

WORKING THE SCENE: DIFFERENT SKILLS

As scenes become more complex and diverse, the range of crime scene personnel is expanding. This section explores two positions that, though always on the edge of scene investigations, have grown markedly over the past decade, reflecting advances in technology and in the growing need for specialists—as opposed to jacks of all trades—in all fields.

One of these positions, though not all that visible on a day-to-day basis, that is rapidly changing and may well become much more prominent in the future, is that of animal handler.

Animal Examiners

In an increasingly technical world, some investigators are turning to more organic methods of addressing crime. Once lumped together under a group description like "K-9 teams" and used almost exclusively for tracking suspects or sniffing out drugs at airports, the role of animal handlers and dogs in law enforcement—and particularly in forensic examination—is widening.

To date, dogs—also thought of as "animal examiners"—are being used for many tasks:

- locating a decedent's remains
- locating objects associated with the decedent

- locating living individuals after avalanche or similar natural accident
- locating lost individuals in rural and urban settings
- supporting and defending law enforcement officers on foot
- scenting and locating accelerants
- scenting and locating drugs
- scenting and locating explosives, including plastic explosives that can't be found by X rays or metal detectors

THE NOSE KNOWS

Dogs have millions more olfactory receptors than humans, but they aren't the best sniffers in the world. Both cats and ferrets have much more discriminating noses, and truffle-hunting pigs have long been acknowledged to be excellent snufflers.

"The problem with sniffer cats is simply that they don't *like* to work," says Michael Deering, trainer. "Pigs don't mind working, but aren't that suitable for a lot of scenes. They get distracted by food. And other things. We tried ferrets, and they're a lot of fun, but they have zip attention spans. Dogs like to work, they respond to the praise stimulus. They're bright and see themselves as people partners. Cats don't, so their noses aren't much use to us unless you happen to be looking for a half-rotted fish and the cat is really hungry."

The problem with animal examiners is, of course, that they can't testify at court and court appearances are part of almost every examiner's life. The notion that a person can testify on behalf of the dog, while interesting, prompted the following from a defense attorney: "I understand you're a dog handler, sir. I'm just wondering if you speak good dog, technical dog, or just pidgin dog?"

Deering laughs at the notion of entering dog testimony through a human translator into the court transcript. "I don't think that's where we should be headed. Dogs and trainers aren't supposed to be court witnesses. We're field-workers who can, by reason of a special skill, *suggest* probable courses for a search that has otherwise gone cold. We don't collect evidence—we find it."

Sniff lineups, in the few studies done, have proven dogs' noses sufficiently discriminating to identify individuals from scent samples. But over what period of time a dog might still be able to make a significant ID remains to be seen. There's no doubt dogs can track—that they can key off of drugs or explosives—but that's only sufficient cause to issue a warrant for a human-led search, not to convict anyone on a major charge.

"But that's not always the goal. Often we just need to narrow the search parameters. A guy throws a bloody knife 'somewhere into that canyon,' and we've got a lot of territory to cover. If a few dogs can cover that terrain quickly, we can move ahead on other fronts. No court is going to contest the fact that a dog can *find* objects. Making the connection to suspect or victim is up to the appropriate examiners—and they can testify on their own behalf!"

Dogs in traditional roles frequently doubled as living weapons—"attack dogs," in some parlance—and needed to be big and strong. A "police dog" immediately evokes images of German shepherds and Doberman pinschers. In the new world of forensic dogs, however, large size can be a disadvantage.

"Try getting a German shepherd down a tight space in a collapsed building!"

One of Deering's best people finders is a miniature pinscher named Dolly.

"Small dogs have a reputation for being rats that bark, but I'm beginning to suspect it's the owners, not the dogs, who are at fault. Dolly here is calm under rough conditions. She can squirm into narrow spaces even smaller than the robot probes can handle, and she knows her job."

Watching Dolly at work, it's obvious that she's serious about it. Dolly is primarily a live subject searcher. She'll walk over corpses and even other living people to find the one subject she's been keyed to find. "We wouldn't take Dolly to a scene with an unknown number of subjects, like an industrial accident or a bomb scene. That's not what she does. She looks for one person based on the scent object we give her. Giblet over there," he points to another miniature pinscher, "he's what we term a 'cadaver dog.' He looks for deceased subjects, bodies, or evidence of a cadaver, and he doesn't care who they are, really. The dogs do very different jobs."

So, is it possible to have a dog search for living subjects without a key object?

"I don't have any dogs trained that way at the moment, but that's not to indicate it's not possible. Other teams in the area certainly perform those tasks. Search dogs in avalanche territories, just as an example, are looking for anyone caught in the snow and aren't being keyed off a scent object. That's the sort of dog-handler team you'd need in the middle of that bomb scene—you'd just have to find some really small Saint Bernards. But a search and rescue situation isn't necessarily a forensic situation. The dog who digs people out of avalanches doesn't have to be trained to watch where he puts his feet in the process. Different dogs, different talents."

Not every dog handler or dog is suitable to work crime scenes. Here are a few basic guidelines:

- Handlers must know their animals well. Dogs pushed when they're tired, hungry, or thirsty may alert just to terminate a search.
- Dogs must be calm and focused. Handlers do more than run alongside galloping hounds—they have to keep the same sort of notes as anyone else on the scene, so their animals must be well-trained in both search and obedience.
- Handlers must understand the legal requirements for search—especially if the trail moves from property to property—and must know when to call in for specific legal advice and support.
- Dogs and handlers must also be able to mark possible

evidence while traveling in such a way that evidence is kept secure for collection. Marking must be nondestructive (no big paint circles around spots where the dog paused), clearly visible, and documented in the search log.

Given a dog and handler that know their jobs well, the limiting factor is the persistence of scent at various scenes. The question always put to handlers is simply "How long can scent last?" What they want to know is how long dogs can be useful in search operations. Studies suggest that this varies widely, depending on the scene involved and the type of search required.

"If you're looking for something that's still there and constantly releasing new scent, like a decaying body, then obviously scent dilution is less of a factor than it would be if an object had been removed from a scene and we were looking for evidence that it had been there at some point in time. The scene might be very exposed, with breezes, and that dissipates scent faster than would be true in a stuffy room where no one lived regularly."

Studies and reports on the abilities of dogs to find cadaver scents, human remains, and live individuals are ongoing, but early results indicate that a year—in a closed building—is not outside the scent range of most dogs. As with people, experience counts for dogs, with age being less of a factor than field time. Cadaver dogs at work in the field have, from time to time, found not only the newly dead cadaver being sought, but also bodies that have lain in well-dug graves for as much as fifty years.

Deering notes, "If the body, or body remains, are still about, a properly trained cadaver dog should be able to alert on any evidence with a sufficiently crisp response to be obviously attempting to draw his handler's attention."

The "crispness" of the response will vary from dog to dog, with some locking up in full point, others circling a scent, and yet others employing a vocal alert by barking. Deering hopes that some standardization in alerts can be developed among handlers and trainers. "A trainer might be running a dozen dogs, with a handler accompanying each one. You can't be walking with all of them, so I think it's very important that dogs be able to inform *any* human partner that they've found something. If we standardize, it'll be easier for everyone."

Human handlers are standardizing their own routines, as well. In previous times, dog handlers usually owned the animals. They would come out to help if asked, but had no knowledge of what did or didn't constitute a legal property search, what were or weren't human bones, or how to keep their animals from destroying a crime scene in the process of finding it. That's no longer true.

Attempting to raise the success rate in their field, modern forensic dog handlers are taking on some serious study time after a hard day of training their animals. Deering and his staff have all completed voluntary course work in those areas of criminalistics that apply to their work, have taken biology and anthropology to help them recognize human remains and possible grave sites, and have as sound an understanding of the legal aspects of search and seizure situations as any law enforcement officer should have.

"It used to be that they'd call us out, point us in the direction they thought most promising, and everyone would scatter off into the search. That's not true anymore. We still take our lead from the investigators, but we're also prepared to suggest a variety of search patterns that might be most efficient in a given scenario. We're generally prepared with our own topographic maps when we arrive, and we help the teams identify any hazards before they head into a situation. If we're searching a dwelling, we can do a lot to reassure people that the search won't be destructive and that the animals are well trained—not about to mess in their homes or anything like that. People are often traumatized enough by events that we want to make sure we don't exacerbate the situation, that we'll do the job and get out."

Recently, Lieutenant Robert Sweetapple has noticed a definite change in the handling of animal searches. "There's a calmness about dog searches that's new, an air of quiet competence on the part of the dogs and their handlers. It makes things easier, smoothes communications, and allows us to bring dogs into situations that wouldn't have been deemed appropriate just a few years ago. When Deering or any of his people turn up, they've got intelligent questions to ask. They want to know what protocol they should follow if remains are found—and they know exactly what we mean when we tell them our preferences. They ask about warrants and the limits of the search area, if they can expect immediate support from us in the form of a walk-along officer or if they should call in, and—if they should need to call in—how much and what can they say on an open-air frequency. They're professionals all the way, and it's getting to be that way in more and more jurisdictions."

In the field (whether that's an actual field or a lineup at an airport), animal handlers must keep accurate logs of events as they unfold. "We aren't trying to replace criminalists, or determine if bones are human, or how someone died when we work a scene. But we try to to make ourselves a little more useful by being intelligent enough about what all those other people do. Speaking the same language eases every situation. From our point of view, knowing that larger bones are more likely to be dragged than smaller ones in most scenes with animal predation means we can make an educated guess about where the rest of a disarticulated body might be and get the dogs going in the right direction. Most handlers have a good instinct about how scent can be carried and how it might affect a dog's ability to search. Little gullies and steep hillsides can redirect or stifle breezes. If the dog seems to be going in circles or working a particular area but finding nothing, a well-trained handler can eye the lay of the land and find a different way for the dog to approach that part of the scene. If we can explain in common-sense terms why we sent dogs left instead of right, that's better for us, for the investigators, and for whoever might have to hear that evidence later."

The kinds of scenes in which dogs are used are growing. Vehicle lineups for drugs have become commonplace, and drug-sniffing dogs can work a string of cars much more efficiently than their human partners. As the use of forensic dogs spreads wider, they're being taught the same seek and locate skills, but their keys are now human and cadaver scents.

Deering and Dolly had a chance to prove their worth in 2000, when the tiny dog was given the jacket of a child snatched in a large grocery store. "The security at the store was excellent, and they had the parking lot shut down within minutes of the mother reporting the kidnapping. When we arrived, she gave us the boy's coat and Dolly and I started walking. She keyed on a green Toyota Corolla about an hour in, and we found the baby in the trunk of the 208th car Dolly had investigated. The abductor had run for it on foot and abandoned the child. He could have come to serious harm in the time it would have taken human officers to search that same area or the same number of cars."

Well-trained dogs, in a variety of terrains and situations, can provide valuable leads to investigators—if, as with any other resource, investigators understand the potentials and the limitations.

Forensic Photographers

Though this book has covered many topics, every section has had one activity in common: forensic photography. Clearly, criminalists and law enforcement officers can be, and frequently are, technically competent photographers. But for most departments, forensic photographers are specialists in a field moving every bit as fast as that of the anthropologist or DNA analyst.

Chloe Pairpoint recalls her early days as a forensic photographer with something approaching nostalgia. "It was so much simpler then. The whole 'a picture is worth a thousand words' thing is still true, but long gone is the idea that 'pictures don't lie.' We're a lot more sophisticated now than when I started in the mid-1950s. We know that pictures can lie, and it's not unusual now for a photographer to be called to the stand and grilled about the preparation of any given image. That never used to be the case."

Digital imaging drew massive attention when *National Geographic* moved one of the Great Pyramids a little closer to the other to improve the scene for their cover, making the general public aware of the possibility that what you see might not really be what you get. Of course, such visual tricks were possible before digital imaging became commonplace. Experienced photographers could work visual magic not only in the camera itself, but also in their darkrooms. The O. J. Simpson case brought the issue of digital scene reconstruction to the fore for public consideration. It was pointed out that the digital image meant to represent the "unknown attacker" could prejudice the jury against Mr. Simpson if the figure were shaded in darker colors; the figure, then, might be interpreted as a "black man," and not an unknown at all. Regardless of any opinion of Mr. Simpson's guilt or innocence, the issue of prejudicing a jury with any exhibit, though especially with visual exhibits, remains legitimate.

Says Pairpoint, "We've always known that a green filter on your camera when taking pictures of bloody scenes will turn that pool of blood almost black on film. Now, for investigators, that's an advantage. They can look at the images and get an excellent view of, for example, splatter patterns. In court, however, it *could* be argued that the photographer was attempting to visually imply that a scene was bloodier than it was. There's a fine line for us all to walk."

As most forensic photographers know, some "tweaking" of the images they take is almost inevitable. In 1936, an on-scene investigator took a picture of a bite mark. But at that time, bite mark evidence wasn't as sophisticated as it would later become and, though the photographer was careful to include a scale in the photo, his camera wasn't completely level in relation to the bite. In field situations, maintaining the perfect plane orientation to evidence can be difficult, especially as the photographer may be very limited in where he or she can stand. Over the years, a variety of aids have been popular. One was a laser sighting system that told the photographer all was well when two dots of light aligned. The 1:1 cameras used even today often have flip-down sticks that touch the body or item to be photographed, ensuring that distance and alignment are correct.

In 1936, however, most photographers were officers that just lined everything up as well as they could by eye. To correct the perspective on the picture, the photographer mounted the negative under the developing light in the darkroom and simply tipped the baseboard, holding the photographic paper to the same angle, and exposed the film. Today, the same effect is possible in digital imaging programs like Photoshop, and is commonly used to take older evidentiary shots and realign them. Have the photos been "altered" or not? The photographer would say no, that the process merely creates a more accurate image. To someone unfamiliar with the problem, however, alteration is simply alteration, not necessarily correction, and can create doubts that what is seen is what was really there.

On one hand, the photographer is expected to create clear images of things that are incredibly difficult to see with the naked eye, like a latent fingerprint. On the other hand, juries don't really trust an "enhanced" print, and lawyers, knowing that, can choose to take issue with any image manipulation.

Chloe Pairpoint is quick to point out that there is no such thing as a "straight print" of any picture. "When I stand in a room, I have to figure out what lens to use. That's pretty basic stuff. On 35mm camera gear, the 50mm lens is probably closest

to what the human eye sees. By that, I mean it covers the same amount of space. It's not a panoramic vista that requires you to turn your head from side to side in order to take in a whole scene. But, what if that 50mm lens won't show the whole room? What if you're standing in a corner and simply can't back up any more? Obviously, you have to get those whole scene pictures, so you switch to a wide-angle lens, probably something like a 20mm or a 30mm. Now, already, I've made a choice about the image I'm going to produce. Have I altered the image? I don't think so."

MONICA GRAFTON

Forensic photographer Monica Grafton lays a strange plastic cube next to a footprint. The cube, completely transparent, contains a compass with an oversized face. Grafton has glued it to the bottom. Along all the cube's outer edges, the standard measuring grids used by her department are taped in place. A square of 18% gray paper takes up another spot. On the opposite side is a colorful arrangement of smaller blocks that are used to help ensure color fidelity. "I got sick of trying to figure out which way was north, or trying to ensure that the print color matched what I remembered before going into court, so I made the Lazy Box. I just drop it into the photo sequence for at least one shot per location or roll and I've got all my color rendition guides, directional guides, and measurement guides in there."

As Monica has discovered, juries, lawyers, and judges are becoming just as likely to question the *how* of a photograph as the *what*.

"It's funny, but I suppose it makes sense that a generation that sees the impossible on their TV sets every night would be more likely to disbelieve what they see with their own eyes. In one sense, modern jurists are incredibly sophisticated. They're more likely to understand the importance of perspective and lighting conditions on a particular exhibit. But, on the other hand, they're so used to being surrounded by computers and other high-tech toys doing things they *don't* understand that they really think there's always **some** way to *spin* what they're being shown. It's a strange combination of savvy and naivete."

It's just that juxtaposition of qualities that led to Grafton's first real quandary as a forensic photographer—and proved that even cops can misunderstand the purpose of her work.

"In the general run of things, I'm usually asked to create exhibits, to take photographs of scenes and items. But, occasionally, I'm presented with other people's photographs and asked to make interpretations, or comment on their authenticity or appropriateness for use in court."

Matters of optics and measurement, like calculating the height of a robber from a video surveillance tape at a corner store, generally go to the crime scene reconstructionists. But in situations where a determination of authenticity must be made, a photographer is usually the first person to examine the questioned items.

"Since digital photos have become so prevalent, people will cast a second look at almost any photograph, but in the Travers case there was good cause to seriously question whether the images about to be submitted to a jury were, in fact, real—were actual representations of actual events."

Lawyers, law enforcement officers, and others associated with the courts long ago realized that human memory is fallible. In a case involving a young woman being beaten by two young men in her own front yard, the three neighbors who came to her assistance and another witness who was busy calling the police managed to variously describe the youth who fled the scene as "tall, black with bleached blonde hair," "tall, Hispanic with a light-colored ski-type hat," "medium height, medium complexion, and medium-brown hair," and, to really confuse the issue, "average height, white, and some sort of colored tints in his hair, like blue or green or something on the tips." When the youth already in custody finally gave up his accomplice, *she* was tall for a woman, of Hispanic background, with hennaed hair that was auburn!

"The woman's defense counsel, naturally, made good use of the discrepancies in eyewitness testimony to cast doubt on whether it was his client at the scene. And the blue tints reported by the woman who called in the altercation to police were referred to time and again as evidence that the observer's word should not be taken into consideration at all."

The scene photographer in that case eventually explained at least some of the mixed descriptions by going back to the scene around dusk, the time of the original attack. In bright sunlight, one part of our eyes is working, but at night, or in low light, a completely different set of visual receptors takes over. At dusk, things are complicated by having both parts—the rods and the cones of the retina—trying to take in information at the same time. Add to that the often quickly changing light and it becomes clear that colors observed at dusk can be seen quite imperfectly. And things you'd think would help—in this case the street lights—can actually skew perception even further. Back on the scene at dusk, the photographer had a woman of similar size and coloring walk through the area several times. She found that the overhead streetlights, a sodium source, cast a nearly purple light that dramatically changed the appearance of the woman's skin and hair.

Still, the blue or green tints were hard to account for—until the photographer went to the booth from which the witness had called police and realized that the tint in the booth's glass, in combination with the overhead streetlight, did indeed wash out red tones and turned any hot spots—such as the natural highlights in hair—blue-green.

"My case was similar in that I was trying to figure out if something in a description and, in this instance, a photograph supporting the description, could have been influenced by some previously unrecognized interaction between human and environmental conditions."

From 8:20 to 2:15 one Saturday night, police were watching and photographing a suspected crack house and a youth named Bruno Pavel in an attempt to link him to dealing at several local raves. The officers followed normal procedure and took numerous still shots of Pavel as he came and went, talked with other people, and, allegedly, handed out drugs in exchange for money. The films were developed, and about noon on Sunday, Bruno Pavel was arrested at his home.

Two days before preliminary hearings, Pavel's defense submitted photos of their own. These digital captures taken from digital video show Bruno Pavel in a different colored sweatshirt than the one he is wearing in the police photos. These images also show several other individuals recognizable from the police photos at a rave twenty miles across town—supposedly at the exact same time police were watching the crack house.

Raves often move, and the police knew that there would have been only so many opportunities for Bruno Pavel to have attended a rave, with those individuals, at that location. It seemed Pavel had a photographic alibi that beat the photographic evidence supplied by police. Police knew Pavel had indeed gone to the rave that night, but the images showed two people in different clothing, and the police evidence itself proved that Pavel had no opportunity to change clothes before arriving at the dance. To top it off, a few frames showed a warehouse clock, putting Pavel in two locations at once.

"Digital video was fairly new at the time but, clearly, one set of images had to be wrong. Or, in the biggest muck-up in ages, the police were watching the wrong guy all evening." Monica finishes up her photos, collects the Lazy Box, and heads back to her car. "And I was the lucky newbie who got to figure out what was really going on."

Back at the lab, while her latest images are being processed, she pulls the files on the other case. "Because I wasn't on staff when the original surveillance photos were taken, I got first run at them and the supporting documentation."

The case notes, the photos, the records of the lab and technician who processed and printed them all appeared to be in order. But, as Monica explains, because the defense was suggesting invented evidence, she examined not just the print history, but also the negatives themselves. They were clean negatives, and the prints she made from them were identical in every respect to those submitted originally as evidence.

"We didn't have things that easy on the defense side, though." Pointing to a rather bulky setup in the corner of the lab, she explains, "That snags a single image from a videotape and prints it out much the way you'd print a digital photo at home. The prints and the original tape can be compared much as a photograph and its film negative can, and the possibility of manipulation of the second image eliminated."

Next, she turns to a computer terminal at the far end of the lab. "That's the same scenario, but for digital video. And it's the digital video that made this particular case so difficult."

With a tiny handheld gizmo (that also happens to play MP3 music files, acts as a cheap Web camera, and can hold your address book and send E-mail) she quickly takes thirty seconds of "film" of me walking about in the lab, then attaches the unit to the computer. As I watch, and in less time than it takes to read this sentence, the video begins streaming across her screen. It isn't great quality, but it's certainly easy enough to identify the place and the subject.

"This is typical of the cameras some of the kids at raves carry—techie but not professional quality by any means. We have better quality gear, but this is just a demo to show you what can be done and, honestly, it's easier to make changes look good when you start with poor-quality images. Everyone assumes any little hitches are just digital artifacts anyway."

What can be done is miraculous. In moments, Monica alters the color of my hair, gives me a scar my twelve-year-old would envy, and turns my sedate blue jeans brilliant pink—and absolutely nothing in the background has changed. Colors elsewhere in the frame are perfect. There's no telltale pixelation of the sort we've seen on Internet hoaxes, no strange shadows where they ought not to be.

"See? It's amazing. And, given time, we could make identical changes to each and every frame. And there's no way to know that this wasn't the way it looked originally. So comparing the prints with the digital video is practically worthless at this level. If the DV is already altered, then the prints are authentic duplicates, but authentic duplicates of a hoax."

Yet, clearly the young man couldn't have been in two places at once, so, if you know the police evidence is clean, then the phony images must be the ones submitted by the defense, right?

"Well, we can *know* that, but if we can't *prove* it, the most likely scenario is that all the photographic evidence will be tarred with the same brush."

And how could you prove that this video was retouched or altered?

"At this point, I couldn't. I had suspicions, but there's simply no way, short of examining images and code at a level considerably beyond a photographer's knowledge, that I could prove them."

But the pictures *were* disallowed.

"Frankly, it was pure luck." A few clicks of the mouse later, she's saving her altered video on the magnetic media in the camera and pulling up the new directory listing from the digital camcorder. Her finger taps the screen. "I did know *one* thing about digital images, whether still shots or video. They're really just computer files. See how each file is dated and timed as it's saved? The camera is basically a portable computer with a miniaturized drive. It records the date and time each file was saved. Defense counsel is used to people asking to see the original negatives, but we asked for the camera and the original magnetic media and started looking at the files and file-saving routines, not the images themselves."

So, in the long run, you never did prove that the video was altered?

"Not from a photographic sense, no, though the supporting evidence certainly suggests that was the case. The clock only appeared in a very brief section of the film, and there's a fuzziness about parts of the image that suggests, but doesn't prove, it was altered. You're not talking a lot of pixels to move a black line from one place to another on the face of a clock seen for just a few seconds. The different-colored shirts were the same shirt, just colorized like the munchkins in *The Wizard of Oz*. A nice touch, which, with a jury, might have cast just enough doubt to get it thrown out this time. I couldn't have proven the colors were altered. I *could* and *did* prove that the digital video submitted was created, or at least saved back to its disk, nearly nine days *after* defense claimed it had been shot."

Don't cases like that make your photographic evidence even more difficult to support in court?

"Perhaps. But as technology moves forward, someone some-where is keeping up with it. And unlike whoever altered the Pavel video, criminalists can share information about new technologies and approach photographic problems from more than one angle."

Forensic Photographer's Kit

Camera gear choice is as individual as the jurisdictions photographers cover, so no list will be definitive. This one covers most situations, with more options opening up if laboratory setups are included.

- Cameras. 35mm, Polaroid equipment, video gear, and digital equipment. Any and all, depending on the local authority's policies and the precedents set by courts in that jurisdiction. Some photographers prefer medium-format cameras because the larger negatives give more detail, so it's not unusual to see these items in addition to the others.
- Normal lens. For 35mm equipment, that's a 50mm lens.
- Wide-angle lens. For 35mm equipment, that's a 20mm or 30mm lens.
- Close-up lens or accessories. This might be a "macro" lens, or it can be an add-on for normal lenses. Some lenses not intended for close-up work can be easily adapted to it with reversing rings, which allow the lens to be turned backward to the camera. Extension tubes and bellows are other options for increasing the image size.
- Filters. These add-ons can make certain colors more prominent, separating evidence from a background color. In the case of UV photography, the Kodak Wratten 18A filter, which blocks the other wavelengths, allows the UV images to be separated from normal shots.
- Electronic flash. Essential at night, but necessary in many

other settings, as well. Two flash units can eliminate that huge shadow often cast by a single unit.

- Flash and remote cords. To allow a single operator to handle all his or her equipment without an assistant, it's very helpful to have long cords to set off flash units or cameras. A remote cord on a camera will also reduce shake, which can degrade images.
- Tripod. Obviously forensic pictures are meant to capture detail. That is made much easier if the camera is steady. Tripods are one way to ensure that shots will be in focus, even in difficult situations.
- Film. Lots of film. Black and white, color, and speciality films in a variety of speeds to match existing conditions.
- Documentation book to record camera settings, filters used, etc., for each frame of film. Should include scene number, date, time, and brief description of the item being photographed.
- Measuring equipment and scales. The scale should include a square or grid to allow for perspective checking.
- Level. Some cameras come equipped with a level, as do some tripods. If none is available on the equipment itself, a small level can help ensure that horizontal alignment is correct, which is especially important in creating images that accurately reflect the angle of inclination of roads and hillsides.
- Gray card. A pre-calibrated cardboard card in 18 percent gray. Used to determine optimum exposure. Also very useful if the auto-exposure sensor in your camera happens to die mid-shoot. A separate light meter is also useful for difficult exposures.
- Index cards, marker. To use in photographs as secondary method of identifying the shot or object photographed. Some departments have preprinted versions, but extras never hurt.
- Flashlight. For finding equipment in the dark, providing extra illumination at the scene, and other functions.

- Alternative lighting sources as anticipated need suggests.
- Extra batteries. At least two changes for each piece of equipment.
- Manuals for all equipment. Though photographers are presumed to know their jobs, technical difficulties can arise and having the manual on hand beats having to run back to the lab for it. Likewise, a small repair kit can prove valuable.

As far as the courts are concerned, general guidelines govern all photographic evidence, regardless of what equipment was used to collect it.

- The photograph must not be designed to *influence* viewer emotion. While scenes of violence are expected to create emotion, the photographer must not attempt to manipulate those emotions. It's one thing for a television crew to focus on the image of a child's doll lying amid a crime scene, but a different issue altogether when crime scene photographers do it. If the doll is there, it will be photographed as part of the scene; but there should be no more importance attached to it than to any other item in the scene shot.
- Objects in forensic photographs must also be relevant to the crime. Pictures of grieving relatives are only relevant in establishing who was or was not at the scene. If there's no legal question about that issue, a shot of grieving relatives is not appropriate.
- Photographs must be free from distortions and must not misrepresent the scene or objects in it.

Each and every photograph considered for submission will be scrutinized on those criteria first, but they may be rejected for other reasons if one side or the other believes the shot to be prejudicial in some way. Photographers are required to record the scene as it appears when they arrive, not as it may have been previously. Chloe Pairpoint's first on-scene tussle with an investigator was over whether or not items already removed from a scene should be reintroduced before overall views were taken.

"A photographer can't take the investigator's word that object X was here or there. Not that we assume they're doing anything wrong, just that memory is imperfect. That's why we're there in the first place. Anything that's out of the scene when the photographer arrives stays out."

Some scenes are transitory by nature. Snow melts, rain washes away trace, and a body can't be left at the scene indefinitely. Within those constraints, however, photographers follow a regular pattern in recording the scene, beginning with a consultation with investigators.

- Consult with all those on-site to determine what views and objects are to be covered, and any special circumstances that might relate to photographing those views. Determine if aerial shots will be required, and arrange for those to be taken at the first opportunity.
- Record the entire outer perimeter of the scene, all entrances and approaches. Begin with wide orientation shots, and continue with closer views of items within that view.
- Mark and protect items or scenes that will require further photographic processing.
- Inside the scene, find a central point and take sufficient views to create a complete, continuous panorama. If the scene is the entire interior of a house, every room must be treated as its own scene and the panorama must include all walls, doors, windows, ceilings, and floor surfaces.
- Ensure that all items to be photographed separately are included in at least one overall view, and preferably in more than one.
- Photograph each individual item twice—once with a measuring device, and once without. Why? Because it was once

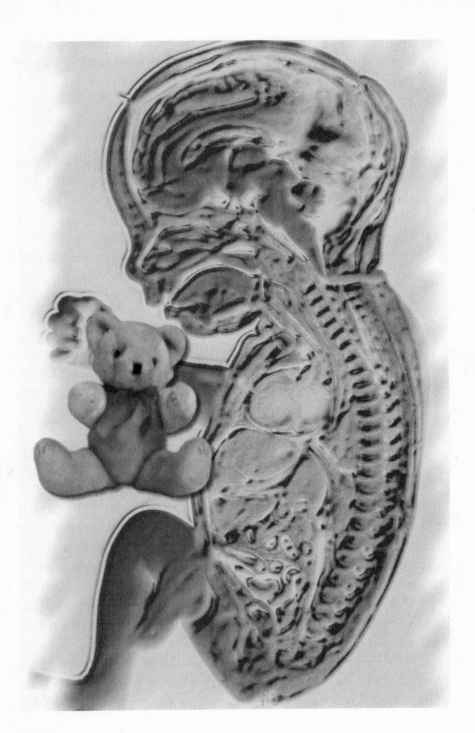

argued that a photographer's ruler hid valuable evidence. Which is also why clear versions of rules and scales can be found.

- Photograph the space left as each item is removed. For example, after all views of the body are taken, photograph the area *under* the body once it has been removed. The medical examiner who finds a mark on the underside of a body and has no shot of the surface under the body is not going to be happy. Only with before and after views can it be proven that an object was *not* under another object.

- Consider the use of alternate photography options. Would the scene benefit from being recorded to videotape? Can instant photos, whether Polaroid or digital images, help investigators right now? Are infrared or ultraviolet images going to reveal something new?

- Take the alternate pictures the scene itself suggests—those requested by investigators, and those that previous experience in the field or in court suggest may be required. Color photos, without filters, are becoming the norm in people shots of domestic violence cases, but capturing the scene in both black and white and color may be prudent in many scenarios. Routine powdered fingerprints reproduce much better in black and white than in color.

For photographers working a scene, it's especially important to consider the individual viewpoint of those involved in the crime. Drivers or passengers seated in a tiny Honda Civic have very different angles of sight than six-foot-tall photographers standing on the elevated side of a highway.

"I learned a great deal from my first hit-and-run scene," recalls Pairpoint. "I'd worked that scene like any other, and it was only when I got in my own car to leave that I realized there was a street sign at one of the corners. I couldn't believe I'd missed that. I could have walked away with a dozen rolls of film taken and not one view of that sign. Needless to say, I hauled myself

back out and started again. And I remembered to get down on my knees so I could see what the five-year-old who was hit actually saw there, as well."

Almost every aspect of crime scene photography is technically demanding. Lighting and capturing a fingerprint in dust require considerable savvy and, as evidence is generally collected, there's no second chance to get it right. The use of fluorescent powders on prints requires yet another approach to capturing the image. Modern lighting (especially fluorescent tubes) bleaches color and can leave a green tone over the entire print, so photographers must know how to adjust film or use filters to return the images to normal color renditions. Lighting of photographs used to present pretty limited options—either natural light and fast films with less detail, or slower film with more detail and bulb flash. Today's photographers have many more choices available, but they must be able to discriminate among them. Light transmitted through a surface—glass, for example—can offer great contrast to prints found on that surface. Side lighting works best for shoe prints in dust, but can hide detail when used on a crumpled fender at a traffic scene. And crime scenes aren't quiet studio locations. People move through them, lighting changes in outdoor scenes, environmental factors like rain play havoc with the scene itself and the equipment. Pairpoint has had film get so cold that it has shattered inside her camera when she advanced the film. "Every time you make a decision at a crime scene, you're altering the view in some way. And in the back of your head, you're already rehearsing the reason you'd have to provide for this choice or that if questions were raised about a shot."

Adding to the confusion is the fact that some photography techniques are *designed* to show features *not* visible to the naked eye. "Photographers seldom testify about techniques, just whether or not they took the photo and what the conditions might have been at the time. But with techniques like UV photography becoming more common, photographers may have to act as expert witnesses, describing the methods to the jury, explaining what is and is not possible with a given technique, or just explaining what they're viewing," Pairpoint said.

Three techniques in particular seem to trip up viewers:

- Night photography
- UV photography
- IR photography

Pictures form on film when light bouncing off objects hits the film and causes a chemical change. Therefore, night, a time when there is normally very little light, requires photographers to use special techniques to create images. One typical answer to the problem is to use a "faster" film—one that is more sensitive to light. These films have their own problems, though, and don't normally deliver the same detail as other films. Someone unaware of the limitations of these films can look at the resulting pictures and think it was badly shot and tend to put less faith in the image's contents. Surveillance work, which can also fall under the umbrella of forensic photography, takes a different approach altogether with the use of "night vision" photographic gear. Once again, these images often appear somewhat out of focus, and certainly less clear than a typical photograph. Before pictures taken by these methods are introduced, some comment on how they were taken should be made. Also, a similar statement should be included with the images when they're returned to investigators.

One other form of night photography that often confuses viewers is the "light painting" method. If images are only formed when light strikes the film, then it's possible for a single person with a single flash—or even a flashlight—to create a shadowless image, even at night. Basically, the photographer opens the shutter and leaves it open, focuses on the object to be lit (often a car or other vehicle), then walks around the object while pop-

ping the flash from all sides or shining the light from all angles. The light bouncing off the object creates one image as long as the shutter is open. With this method, the top, the sides, and even the space under the vehicle (which is normally shadowed in flash pictures) are all illuminated at once. This isn't the usual way to light a night scene, but it can be very useful if a photographer is working alone or doesn't have a multitude of flashes that can be rigged to fire at once. The images obtained don't look like regular photos, though, and as people subconsciously notice things like missing shadows where there ought to be some, the circumstances of these pictures should be explained, as well. Otherwise, the viewer is left with the nagging sense that something isn't right, but can't necessarily vocalize that doubt.

Despite relying on standard photo equipment, UV, or ultraviolet, filming operates on a level completely different from traditional film. Often used in cases where long-term abuse is suspected, UV images can literally see *beneath* the skin. UV light, which causes us to tan, penetrates skin more deeply than other forms of light, illuminating deep bruising and the remains of bruises that can no longer be seen on the surface. This form of photography captures marks under scars, burns, or tattoos, and even makes visible the underlying bruising of bite marks that have since faded on the surface. UV light also records the natural luminescence of many biological fluids.

Of course, any thinking individual presented with both UV and normal images taken at the same time—one showing distinct bruising and the other showing nothing at all unusual—is going to ask why. The forensic photographer must therefore be prepared to testify as an expert on how the images were collected and what science makes them possible. The investigators, medical personnel, and lawyers are the ones who will present evidence on how the person came to be bruised, not the photographer.

UV Photographic Technique

From a technical standpoint, UV pictures present two immediate difficulties. First, the photographer is attempting to take pictures of something he or she can't see, something under the skin. An entire battery of photos—enough to cover the entire skin—must be taken to ensure that no evidence is missed. Secondly, filters like the one used in this type of photography can interfere with auto-focusing cameras, so it's necessary to either keep the individual at an exact, premeasured, and focused distance from the camera or to constantly take off and replace the filter between each shot. Either situation can be tiring and traumatizing for the person being photographed, so photographers dealing with live subjects must be very sensitive to the person's reactions.

On a related note, if the photos are of intimate body parts and the subject is alive, the number of people at the shoot should be severely limited to spare the individual further emotional trauma, while still ensuring that the images obtained satisfy all legal requirements.

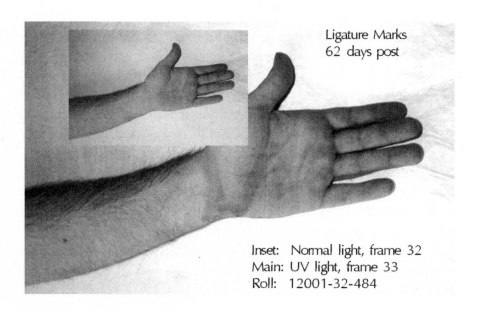

Ligature Marks
62 days post

Inset: Normal light, frame 32
Main: UV light, frame 33
Roll: 12001-32-484

Actually taking the images is, fortunately, a straightforward procedure.

UV light is found in normal light, so most camera flashes produce UV light as a matter of course. A nice, strong flash will provide all the UV light needed and a second, handheld flash will allow photographers to move the light instead of the subject and reduce refocusing time significantly. Most film will also record UV light, so no special films are needed. Black and white film works very well and is standard in most photographer's kits. Normal pictures (used for comparison, as well as to orient the UV photos to come), as well as the UV photos themselves, can all be shot on the same roll of film. No special handling is needed during development, either—another plus.

What is required to filter out all light but the UV is a Kodak Wratten 18A filter. It doesn't look like much, and can be slipped into place on a camera in just seconds, but it eliminates all surface images that would register with regular white light. If there's a bruise under a tattoo, the UV image will show the bruise, but the tattoo will disappear.

To get good UV photographs, then, follow these steps:

- First arrange the camera on the tripod and focus on the space to be occupied by the subject.
- Set up general lighting and attach a handheld flash.
- Provide something a living subject can either sit on or hold to help him or her remain in one position while the shoot progresses.
- Ensure appropriate privacy for the subject.
- Take establishing shots.
- Attach the filter.
- Take the UV shots.

One episode of *C.S.I.*, "Blood Drops," illustrates not only the technique photographers need to adopt to record these images (and shows excellent examples of the type of images delivered from a UV shoot) but also the sensitivity a photographer should develop when taking pictures of living individuals.

UV photographs are becoming standard evidence in cases of individual abuse and offer investigators one special advantage. A bruise can heal on the surface in as little as five days, but bruising beneath the surface—bruising that UV photographs make visible—can be evident for as long as five months or more after the injury. In cases of child and elder abuse (abuse that is typically long-term), these photographs can provide a history the individual may not be able to provide verbally. Through UV photography, rape victims reluctant to come forward at the time of an attack can still provide investigators with evidence that can help catch their assailants.

The IR (infrared) spectrum occurs at the end of visible light opposite from ultraviolet. To picture light that can't be seen with the naked eye, think of a rainbow and add two clear bands—one above the red edge of the bow and one below the blue-violet edge. The one above the red is infrared, the one below the blue-violet edge is ultraviolet. Collecting the UV images isn't difficult, but collecting images in the IR range presents very different technical challenges and creates images that even the most unobservant person would recognize as being not quite normal.

IR images are captured on special films that can be difficult to use at a crime scene. First, such films must be kept refrigerated during storage, but must come back to room temperature before being used. Quite tricky in the field. Secondly, they're incredibly sensitive. Even in its cartridge film can't be exposed to ordinary light and cameras must be loaded and unloaded in complete darkness. More trickiness. Like UV photography, IR photographs require a filter for the lens—this time a Kodak Wratten 87 or something similar. The images produced are decidedly surreal. Skies are black, as is most water. The new growth on plants is brilliant white. Color IR film does exist, but

the colors created on the film have little to do with reality; plus, even more filters are necessary to bring the final image even close to what a human eye sees. Not surprisingly, these images produce a lot of frowns when they're introduced.

Why go through the hassle?

Because IR films can show us what no other film does.

In 1985, a man struck and killed a pair of teenagers walking to school. The driver left the scene and no one saw the incident. After much plodding investigative work, officers developed a suspect, but couldn't find any evidence at the scene or on the car to tie the two together. And in the three months since the incident, they strongly suspected the car had been painted, which further complicated the case. Another month of searching turned up no one who would admit repainting a 1984 Ford Cougar. The dealers who came to look at the car couldn't state with any certainty that the present paint wasn't the factory original—it certainly looked original. The officers were almost ready to give up.

Paula Hogan, a forensic photographer, while in to discuss a completely different case, overheard investigators cursing the lack of leads in the hit-and-run. She'd never used IR photography on a forensic level, but knew its properties from some artwork she'd done at college. "Without really thinking, I sort of suggested that perhaps IR film could pick up any alterations to the paint without actually damaging the vehicle."

The officers jumped at the idea and quickly hauled the vehicle in for her to work on. Within the hour, the photos were taken. Within the day, the prints were back, showing that the surface that appeared so uniform hid two paints that reacted quite differently in the IR spectrum. Within the week, they had a confession.

More typical uses of IR photography include the identification of altered documents (in surveillance work where normal light is unavailable) and in the documentation of some stains and powder burns.

DIGITAL IMAGES IN COURT

National Geographic's altered-image blunder notwithstanding, digital images are gaining acceptance in most courtrooms, as long as the two standard tests applied to all evidence are fulfilled: the image must be relevant to the case, and the image submitted must be open to authentication. Authentication requires that the photographer be known and capable of saying "Yes, I took this image." Questions about how the image came to be taken or what might have been done to it before arriving in court as evidence are handled like any other evidence. Under most jurisdictions, any alteration to the image would be noted at the time the image was offered as an exhibit, and it would be expected that the raw data would be available for review. For photographers and the courts, the real issue has become the security of digital images. Officers obviously need to be able to retrieve images for use in an investigation but cannot be allowed to alter them in any way. Read-only computer media has, therefore, become the preferred storage for forensic images, whether of still pictures or of digital video sections.

Changing image formats can inadvertently result in lost data, so photographers generally insist that the final image come from the format originally used to record the image. If an image was originally a .tiff file, it should not be converted to a .jpg or other format. And *only* the original file format should be used to print images for submission.

Clearly, even in a reasonably large and diverse unit, duties overlap. Photo technicians aren't the only people capturing images at a crime scene. Field personnel and laboratory investigators also cross the crime scene tape to put evidence in context. The schematics, diagrams, reports, and exhibits finally presented to law enforcement, judges, and juries—all the exhaustive documentation involved in a major case—may pass through the care of hundreds of hands. The goal for all is the same—to successfully clear cases. Which brings us to one of the nearly universal duties of all criminal investigators: court appearances.

"Court testimony is a classic love-hate situation," laughs criminalist Bill Bonner, who has been sworn in as an expert witness in over 2,500 cases. "On one hand, you're delighted to get *to* court. It means the evidence you've collected—the casework done by the detectives, the piles of paperwork—*meant* something!"

Even with modern criminal investigation techniques and dedicated personnel, huge percentages of cases in first-world countries aren't being solved. In Las Vegas, break and entry cases may not even warrant a call-out for forensic personnel, and the percentages of burglary cases cleared, or of goods actually recovered, is disappointingly small.

"So, just getting a case to court is something of an accomplishment in some situations." Bonner frowns as he runs a finger over his personal cold files, the unsolved cases to which he still returns from time to time. "There's a high standard to meet in taking a bunch of disjoint evidence, statements, and street knowledge and turning that into an arrest. I've seen cases that look great on paper fall apart and seen juries convict on evidence so suspect I wouldn't have bothered to bring it to court, but those are the oddballs. In the majority of situations, your case has to be tight before you walk into a courtroom with it." He grins. "Where everyone will do their best to make you look like an idiot!"

In Bonner's opinion, the single factor most likely to keep otherwise qualified people out of the field is courtroom interrogation. "You spend hours working on this evidence, days second-guessing yourself, asking if there was more you could have dragged out of it, something you missed—all the while adhering to the strictest standards of policy and procedure—just to have someone say, 'Are you sure you didn't spill your salad dressing on this sample, sir?' "

Traditional crime novels and movies seldom relate the hours spent writing reports or preparing for trials, or the long hours waiting about in corridors outside the courtroom. "It's boring, so you can't blame fiction writers for glossing over it, but it is a huge part of what we do. Depending on your area of expertise and your geographic region, you might spend up to half your time on things like that, but it would make lousy reading. I think every criminalist probably watched *Quincy* at some point, and among all those episodes, I can only think of half a dozen when the story followed him into the courtroom—and I never once saw him write a report!"

More contemporary depictions of criminal investigations, like *Law and Order, Crossing Jordan,* and *Blue Murder,* broadened audience understanding of the difficulties facing "friends of the court," but it may have been the all too real coverage of the O. J. Simpson trial that really brought home the less glamorous side of forensic science.

A quick glance through the Simpson transcript reveals nearly a hundred instances of direct questioning on the *quality* of lab and crime scene work during that investigation. "Clearly, the Simpson case was atypical in any number of ways," says Bonner, "But I think it was the first time the general public—which doesn't usually see the multitude of events that happen before trial described in any detail—had the opportunity to appreciate the mass of paperwork, the importance of rigorously controlling the chain of evidence for every exhibit, the time constraints that scientists often struggle with, and the browbeating an individual must take in court just for the opportunity to present evidence to a jury."

Carly Vickers, a law professor whose speciality is jury dynamics, remains amazed by the composure demanded of forensic scientists. "They have the already near impossible task of breaking down complex scientific principles for a lay audience made infinitely more difficult by a battery of lawyers questioning their competence, their knowledge, and their motives! While most of us have to account for our actions, few people have to prove their ability so completely day in and day out."

Bonner couldn't agree more. "Everyone is fallible, accidents can happen, bizarre instances abound in most cases. But I've yet to meet the investigator who doesn't include those few, rare instances in their reports, or who won't freely admit that there are sources of error in their methodology, or that the field is always moving forward.

"When you're asked at trial—and every decent lawyer will ask—if there aren't more modern, or more discriminating, or more comprehensive tests that might have been undertaken, there's not a criminalist worth their salt who wouldn't answer, 'Of course!' There's always more we *could* do if we had unlimited budgets, the latest gear, and all the time in the world. But that there's more we could do doesn't make the testing we have already done less appropriate or our results less certain. Most criminalists have come to accept that not only do they have to defend their own abilities, but they have to defend the entire field of forensic science every time they get on the stand."

Vickers agrees, adding, "Juries don't like science, they often feel incompetent to understand the concepts. There are *always* new concepts. And they're so busy worrying that they aren't going to get it that they aren't really listening, so it's a self-fulfilling prophecy." Vickers has interviewed hundreds of ex-jurors. "It isn't that they aren't trying, but that they're trying too hard. Jurists are generally very civic-minded. They really want to do the right thing. They're usually terrified of misunderstanding, of misusing information in a way that will change someone's entire life, of making the wrong decision. It's frightening for them. Outside the pressure of the case, they understand the concepts with no more effort than they'd expend learning to operate a new piece of equipment or programming their VCR. Inside that courtroom, though, the pressure is intense and they unconsciously blame the criminalists for asking them to understand."

Bonner doesn't blame the jurists or the lawyers, though. "We all have our parts to play in the system. If we weren't challenged constantly, it's possible the standards wouldn't be as high and that sloppy work might sneak in. Still, for the new investigator, it's daunting. They have to learn a lot on the job. No classroom could cover all the permutations of real life."

As a professor in criminalistics, Patrick Doyle watches lots of new investigators head into the field. "In a lot of ways, it appears a thankless sort of work. Long hours under difficult conditions, with call-outs probably more common on holidays than not, all to get to court and have all your hard work—not to mention your basic common sense and honesty—questioned. It can make you wonder why anyone would go into the field!

"Most people have a real desire to use their talents to 'make a difference.' They could undoubtedly find less stressful positions outside law enforcement. Take Carl Kender as an example. He worked in questioned documents for just over twenty years, earning about the equivalent of $32,000 a year. When he retired in 1998, he went into business privately and now does a little work for several autograph sellers in and around D.C. He makes twice as much working half the hours! He comes to speak to a lot of my students and, when they invariably ask why he didn't go private years ago, he has one answer—'Luke.'

"That's the name of the victim in the first case he was assigned. It was a kidnapping, and things went badly on most fronts. The only lead was a demand note. They found the boy. He was four, abandoned in a barn in the middle of winter. The farm was unoccupied, the nearest neighbors miles away, and it was cold. There's not a person who worked that case that believes the boy could have survived much longer. Most careers start less dramatically, but I think the hope of being able to 'make a difference' is a powerful incentive for most new investigators. It sure isn't the money!"

AFTERWORD

Researching this project put me in direct contact with nearly two hundred individuals in more than thirty police forces, private labs, public forensic labs, teaching universities, hospital research facilities, and even private individuals who've turned their years of experience as public servants into thriving independent practices. Of the 428 people asked to contribute their insights or most interesting cases, only one declined—he was involved in a major crimes case and didn't feel he would be available for follow-up interviews. The remaining 427 turned up old exhibits, passed along contact information for yet more interview subjects, and, regardless of their personal schedules, never failed to call back on time.

You can't fail to be impressed by their stamina, their unmitigated enthusiasm for their individual fields, their dedication to remaining current in sciences that change daily. Yet, in a discipline that surprises every day, they seem most surprised by the public's fascination with their work!

"It amazes me that people ever ask what I did at work today," says trace analyst Bernie Cohen. "I mean, I know why *I* get all worked up about this fiber or that plant pollen or even the hairs found on a suspect's jacket—I just don't get why anyone else would. I mean, we're talking science-nerd central!"

Criminalist Pat Yakama is more sanguine about the public's fascination. "Well, it's like a little mystery puzzle every time, isn't it? Who doesn't like puzzles? Sherlock Holmes, or *Quincy*, or that new one—*C.S.I.*—they're all mysteries and people love to read or watch, see if they can't beat the investigators to the finish line, figure it out first." He taps his "cold cases," the ones he's yet to solve. "I know investigators with file cabinets full of cases in their basements. It's addictive, knowing that there's an answer somewhere, but that you can't just flip to the back of the book and find it."

"For me, I try to keep it as academic as I can," says Beatrice Baker, who's been involved in DNA analysis for just over fourteen years. "If I let myself get wrapped up in the cases, instead of the immediate task, I can be as bad a lookie-loo as anyone else. The O. J. Simpson trial just about drove me nuts. It's not that big a field that you don't know, if only by name, at least some of the people involved in a major case like that. So, especially when they were presenting evidence in your field, you couldn't keep your ears from pricking up. Next thing we knew, someone brought a portable TV into the lunch room. We were way worse than all those women they showed in the laundromats!" She laughs. "I can see now how people get hooked on police shows and murder mysteries—you just got to *know* how it all turns out. I guess crime is, in the broadest sense, human theater. Everyone wants to know if the butler really did it."

Beatrice's partner and workmate, Calum Price, agrees. "There's such satisfaction in taking in the evidence, doing the work, and knowing you've found the right answer. You might not be able to solve the bigger puzzle of the crime itself every time, but you can solve your bit, and even that's a real high. I don't think it matters if you work in the field or not. It's just terribly satisfying to take all that uncertainty, the chaos of a crime scene, and put it all back together in one coherent story. Ya gotta love that!"

With shows like *C.S.I.*, the various incarnations of *Law and Order*, and even the considerably unorthodox investigators of *The X-Files* glomming up millions of viewers each week, and with forensic investigators as different as Patricia Cornwell's

Kay Scarpetta and Ellis Peter's Brother Cadfael regularly claiming readers' attentions, it seems Price is right. The thrill of the hunt is universal.

Tigh Kelly's online mystery fanzine, *ClueQuest*, features a real-world case in each issue, and it's his readers' favorite section. "We set up the evidence, but don't reveal the perpetrator until the next issue. We were three days late sending out one mailing and you wouldn't believe the e-mail that hit our mailbox asking for the solution! Everyone, not just our readership, becomes invested in a good story and, despite the human misery involved, most people treat even real crimes as a sort of performance art. They can't change the events that have already happened, so I don't think it's particularly morbid, but there's a definite lure to an unsolved mystery.

"Back in high school, our math instructor put up an equation, a problem really, on the blackboard and told us there was no solution to it. I don't, today, remember one specific thing we learned that semester, but I'll never forget the energy we all heaped on that question—and the frustration we felt knowing that the teacher wasn't simply going to walk into next class and tell us the right answer! Cruel man."

John Bandoaloa, one of Kelly's most popular writers, figures he knows why mysteries are so infectious. "People love closure. You can write the most intriguing story, people it with vivacious characters, twist the hell out of the plot, but if you can't provide a satisfying ending—good, bad, it doesn't matter as long as you tie up the loose ends and play fair with the reader, no stupid stuff—that story is a waste of paper. No one thinks in vignettes. Give us any scene, or any glimpse of a real crime, and we have to cast it in story form, which means we want to know what went before and what happens next. We aren't happy critters if we can't say this happened, and then this, and so on. Our brains are hard-wired to narrative."

"Mysteries satisfy that need better than any other drama," adds Kelly. "Mysteries are set up to answer one question at a time. Another form of drama might ask us to understand the entire arc of a character's life, or draw some lesson from that life, or even to put ourselves in the character's place. Mystery doesn't. It's not simple in content, but it is in structure. It asks us to solve one problem. If we do that, preferably before the end, we've accomplished something, reached a clear-cut goal. You just can't beat a good mystery for pure satisfaction."

APPENDIX A

THE MANY FACES
OF FORENSIC INVESTIGATION

While titles vary from jurisdiction to jurisdiction, the basic classifications of duties frequently fall into similar groupings of responsibility. The Las Vegas Metropolitan Police Department, which has a midsized forensics unit that, with a wide variety of crimes, does not have C.S.I. IIIs, but it has a well-staffed unit whose duty breakdowns can provide insight into the challenges—and rewards—of many forensic science/crime scene positions.

Evidence Custodian

Performs a variety of technical duties to ensure that evidence is properly processed, stored, protected, and delivered, or to release and dispose of cleared property.

Major Tasks

- processes, transports, and stores evidence and property
- releases property to appropriate personnel, delivers evidence for laboratory analysis; packs property for shipment; prepares cleared property for final disposition; destroys cleared narcotics, weapons, and biohazardous material
- is computer literate, familiar with related peripherals; creates and updates evidence records/chain of custody records; retrieves information from computer applications; prints and delivers reports on evidence/property

- photographs evidence, keeping appropriate documentation of activities
- registers firearms in property
- conducts quality control audits on evidence vault inventories

Collateral Functions

- verifies the amounts of currency impounds
- makes bank deposits; purchases cashier checks
- operates a forklift if required
- optionally, qualifies with a department firearm

Qualifications

Required Knowledge

- relevant principles and procedures for the retrieval, categorization, preservation, or disposal of evidence
- relevant methods and techniques of quality control
- familiarity with relevant laws, codes, and regulations at all levels
- principles and procedures of recordkeeping, computerized or otherwise
- familiarity with modern office procedures, methods, and equipment
- familiarity with modern inventory/warehouse equipment and methods

Required Skills

- store and dispose of evidence
- maintain accurate, correct records
- work independently, without immediate supervision
- interact with the general public
- communicate clearly and concisely, orally and in writing

- establish and maintain cooperative professional relationships
- maintain physical and mental conditions necessary for the performance of assigned duties and responsibilities

Experience and Training Requirements

- Experience: one year F/T experience in law enforcement support, or storage and retrieval of inventory, with a typing speed of 25 wpm
- Training: possession of a high school diploma or GED

Licensing and/or Certification

- valid driver's license of appropriate class

Working Conditions

- Environmental conditions: various, which may include confined/small spaces; heights on ladders; possibility of temperature extremes, loud noises, and airborne particulates at either the warehousing facilities, on the scene, or during evidence transport
- Obnoxious or demanding conditions while handling evidence items like human tissues and fluids or weapons allegedly used to commit crimes
- Physical conditions: A necessary ability to perform moderate lifting, operate motorized vehicles, climb ladders, stand or sit for extended periods

Crime Scene Analyst I/II

Responds to crime scenes, performing wide variety of investigative tasks to document the crime including—but not limited to—on-scene photography, recovery of physical evidence, and processing of latent fingerprints

CSA I: The entry/trainee level of the CSA series. Categorized by the performance of the more routine tasks/duties, responding to less complicated crime scenes, and assisting level II staff. New employees may have limited directly related work experience or none at all.

CSA II: The advanced/journey level of the CSA series. Categorized by the performance of the full range of duties, responding to more complex crime investigations, and assisting at autopsies. CSA IIs receive only occasional instruction or assistance as new or unusual situations arise. They are proficient in all policies of, and procedures within, the work unit.

Major Tasks (Levels I/II)

- responds to/investigates crime scenes
- ensures adherence to department safety precautions
- performs documentation of the scene by photographing the scene, fingerprints, and other evidence
- performs documentation of the scene by sketches/diagrams
- prepares accurate reports and documentation of crime scene
- processes (by powder or chemistry) latent fingerprints; performs and submits fingerprint comparisons; classifies fingerprints as appropriate
- collects, preserves, and appropriately packages crime scene evidence
- collects, unloads, and impounds firearms
- testifies in court with expert status

Collateral Functions

- maintain current knowledge of trends and innovations in the field
- qualify with department weapon

Qualifications (by Level)

Required Knowledge (CSA I):

- understanding of basic standard safety strategies
- basic understanding of chemistry, biology, related physical sciences
- basic understanding of drawing and sketching techniques
- basic photographic principles
- familiarity with pertinent codes and regulations at all levels

Required Knowledge (CSA II):

- all knowledge required of CSA I
- advanced latent fingerprint processing including magna-brush
- broad overview of total crime scene investigation program, including operations, services, and activities available
- all departmental policies and procedures
- statutes pertaining to crime scene investigations, fingerprinting, photography, and other tasks of the work unit

Required Skills (CSA I)

- learn the operations, services, and activities of a crime scene investigation program
- learn powder and chemical latent fingerprint processing
- learn departmental policies and procedures
- learn statutes pertaining to crime scene investigations, fingerprinting, and photography
- safely operate and handle firearms
- communicate clearly and concisely, both orally and in writing
- establish and maintain effective working relationships with

those contacted in the course of work, including department officials and the general public

- maintain physical condition appropriate to the performance of assigned duties and responsibilities, which may include the following: walking, standing or sitting for extended periods of time, operating assigned equipment
- maintain effective audiovisual discrimination and perception required for making observations, communicating with others, reading and writing, operating assigned equipment and vehicles
- maintain mental capacity that allows for making sound decisions, demonstrating intellectual capabilities

Required Skills (CSA II)

- organize and prioritize the crime scene investigation
- use and operate material and equipment used in crime investigations

Experience and Training Requirements

- CSA I/II Training: equivalent to an associate arts degree from a community college with major course work in criminal justice, forensic science, physical science, or a related field, including specialized training in crime scene investigation
- CSA II Experience: two years of crime scene investigation experience as a CSA I with LVMPD

Licensing and/or Certification (I/II)

- possession of, or ability to obtain within one year of hire as a CSA I, Forensic Science Certificate issued by the American Institute of Applied Science
- possession of, or ability to obtain, a valid driver's license

Working Conditions

- Environmenal conditions: office environment and travel from site to site; exposure to human body fluids; exposure to hazardous chemicals; exposure to inclement weather conditions
- Physical conditions: essential and marginal functions may require maintaining a physical condition suitable for moderate or light lifting; climbing, bending, or stooping; crawling in confined spaces; heights; standing for prolonged periods of time

Criminalist I/II

Performs various scientific analyses in a laboratory setting on physical evidence to deliver scientific consultation; to interpret and form conclusions from test results; to document such interpretations and conclusions in a variety of reports; and to testify as an expert at court proceedings.

Criminalist I   The entry/trainee level class of the criminalist series. Categorized by the performance of the more routine tasks/duties, requiring less breadth or knowledge/experience, in the forensic laboratory setting. New employees may have limited directly related work experience or none at all.

Criminalist II   The advanced/journey-level class of the criminalist series. Distinguished from the Criminalist I by the performance a broad variety of duties in the forensic laboratory setting, including performance and development of complex laboratory tests and specialization in one or more specific forensic areas. Criminalist IIs work independently, only receiving assistance/instruction in unusual situations. They are fully cognizant of operating procedures and policies within the work unit.

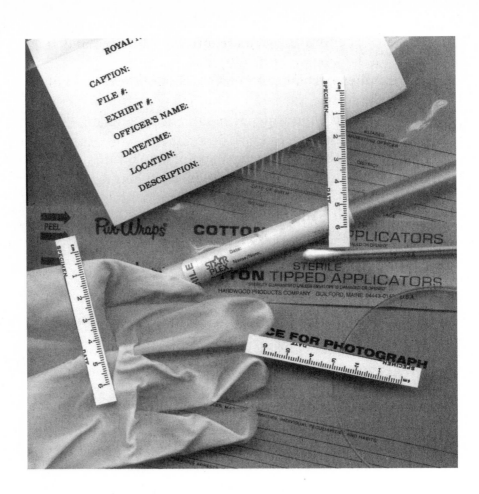

Major Tasks (Levels I/II)

- collects, preserves, and maintains the integrity of laboratory evidence; checks the accuracy of incoming information; assesses evidence to determine proper testing techniques; documents the chain of custody
- performs chemical, comparative, instrumental, microscopic, serological, toxicological, trace testing procedures on various materials within the laboratory
- isolates and separates analytes from various matrices, including solid dose samples and biological fluids

- prepares chemical reagents, controls, solutions, and standards under laboratory conditions
- is computer literate and familiar with various specialized scientific software
- is familiar with a variety of laboratory equipment (i.e., chromatographs, breathalyzers, spectrophotometers, immuno-assay and electrophoresis instruments) and can operate, calibrate, troubleshoot, and perform minor repairs to them
- interprets test results, collects statistical data, prepares scientific reports and affidavits
- projects a professional image while testifying to scientific principles and testing results
- trains law enforcement personnel
- provides opinion on investigative leads and testing reports

Collateral Functions

- maintains current knowledge of trends in the field of forensic science
- actively conducts independent research to evaluate new methodology and, if desirable, initiates new procedures
- provides scientific assistance to attorneys, state lawmakers, and regulatory agencies

Qualifications (by Level)

Required Knowledge (Criminalist I)

- theoretical and analytical principles of natural and physical sciences (including organic, inorganic, and physical chemistry; biochemistry; biology; and other applicable fields and subdisciplines)
- statistics, higher mathematical principles
- standards of laboratory testing procedures and methods
- familiarity with equipment and supplies used in a chemistry laboratory

- WHIMIS system, proper laboratory methodology, safety procedures in the areas of chemistry, toxins, and biological samples

Required Knowledge (Criminalist II)

- all knowledge required of Criminalist I
- broader evidence collection and preservation procedures
- advanced forensic scientific theory and practice
- specialization in one or more areas, such as alcohol analysis, clandestine laboratory response, serology, toxicology, or trace analysis
- knowledge of codes, laws, statutes, and regulations pertaining to forensic science at all levels of government
- overviews of the policies, procedures, rules, and regulations governing a forensic laboratory
- operational methodology and techniques applicable to the use of forensic laboratory equipment

Required Skills (Criminalist I)

- learn procedures for evidence collection/preservation
- learn theory and principles for handling controlled substances
- learn statutes, laws, and codes at all levels of government pertaining to forensic science
- safely handle chemical and biological hazards
- learn policies, procedures, rules, and regulations governing a forensic laboratory
- learn methodology and techniques of forensic laboratory equipment
- perform various scientific tests and analyses
- recognize anomalies in evidence and test results, formulate hypotheses, and take appropriate actions
- be computer literate

- communicate clearly and concisely, both orally and in writing
- establish and maintain effective working relationships with those contacted in the course of work
- maintain effective audiovisual discrimination and perception needed for making observations, communicating with others, reading and writing, operating assigned equipment
- maintain mental capacity that allows the capability of making sound decisions, demonstrating intellectual capabilities
- meet age requirements for handling alcohol and controlled substances in a laboratory in the state of Nevada

Required Skills (Criminalist II)

- all skills required of Criminalist I
- troubleshoot, perform, and supervise minor repairs on laboratory instruments and equipment
- work independently in the absence of supervision
- review and evaluate new and existing information and analytical techniques for implementation into laboratory protocol
- provide training to Criminalist Is, other department personnel, law enforcement personnel, and attorneys, when and as needed

Experience and Training Requirements

- Criminalist I/II Training: Equivalent to a bachelor's degree from an accredited college or university with major course work in criminalistics, forensic science, chemistry, biology, or a related field, including 24 semester hours of chemistry
- Criminalist II Experience: Three years of professional research and practical experience working in a forensic laboratory, including court testimony as an expert witness

Working Conditions

- Environmental conditions: laboratory environment; exposure to potentially hazardous chemicals; exposure to human body fluids
- Physical conditions: may require maintaining physical condition necessary for light lifting and standing for prolonged periods of time

Document Examiner

Examines documents and document-related evidence within the environment of the scientific laboratory; interprets results; forms conclusions; testifies at court as an expert witness.

Major Tasks

- collects, preserves, and maintains evidence integrity in the laboratory setting, ensures accuracy of incoming reports/ information; accesses evidence

- ensures chain of custody is maintained
- assesses evidence and determines appropriate testing schedules
- performs chemical, comparative, instrumental, and microscopic tests
- examines documents and related evidence including, but not limited to, handwriting, printing, and signatures
- performs various tests to, and interprets the examination of, indented writing, mechanical impression, stroke sequence, writing interlineation
- evaluates changes, alterations, erasures, and obliterations to documents
- is computer literate and familiar with appropriate software
- can operate, troubleshoot, and perform minor repairs to laboratory instruments including, but not limited to, microscopes, Electrostatic Detection Aparatus (ESDA), and photographic equipment
- examines papers, copier products, computer printers and their documents, establishes date of origin
- uses alternative light sources (i.e., IR/UV—infrared and ultraviolet) to determine similarity of inks/toners
- compares mechanically printed, photocopied, and electronically produced documents, interprets and documents results
- analyzes and interprets test results, collects statistical data, produces scientific reports and affidavits
- projects a professional image while testifying to scientific principles and interpretation of results
- trains law enforcement personnel and provides consultation regarding investigative leads and analytical results

Collateral Functions

- stays current in new trends and innovations in the field of forensic document examination; conducts independent

research, and evaluates new procedures and methods, recommending new procedures and updating existing ones
- provides scientific assistance to attorneys, state lawmakers, and regulatory agencies

Qualifications

Required Knowledge

- laboratory evidence collection and preservation procedures
- theory and principles of forensic science
- handwriting/printing; photography; alterations of documents; business machines; and ink and paper analysis
- federal, state, and local laws, codes, and regulations pertaining to forensic science
- policies, procedures, rules, and regulations governing a forensic laboratory
- operational methods and techniques of forensic laboratory instruments and equipment

Required Skills

- operate, troubleshoot, and perform minor repairs on laboratory instruments and equipment
- work independently with no supervision
- review and evaluate new/existing information and analytical techniques for possible implementation into laboratory protocol
- recognize anomalies, formulate hypotheses, and take appropriate action
- provide training to personnel, law enforcement, and others as needed
- communicate clearly and concisely, both orally and in writing
- establish and maintain cooperative working relationships with those contacted in the course of work

- maintain mental capacity that allows the capability of making sound decisions; demonstrate intellectual capabilities
- maintains physical condition appropriate to the performance of assigned duties and responsibilities, which may include the following: walking, standing, or sitting for extended periods of time; operating computers and related office equipment
- maintain effective audiovisual discrimination and perception needed for making observations; communicating with others; reading and writing; operating assigned office equipment

Experience and Training Requirements

- Experience: three years professional research and practical experience as a questioned document examiner in a forensic laboratory
- Training: equivalent to a bachelor's degree from an accredited college or university with major course work in criminalistics, forensic science, chemistry, biology, or a related field

Licensing and/or Certification:

- possession of, or ability to obtain, an appropriate, valid driver's license
- possession of, or ability to obtain, a valid Forensic Document Examiner Certification issued by the American Board of Forensic Document Examination

Firearms/Tool Mark Examiner

Performs scientific and laboratory analyses on firearms and tool-mark evidence; interprets test results and forms conclusions; prepares reports; and testifies in court as an expert witness.

Major Tasks

- collects, preserves, and maintains integrity of evidence in the laboratory; ensures accuracy of information received; accesses evidence; maintains chain of custody
- determines proper testing techniques and performs chemical, instrumental, microscopic, photographic, and comparative laboratory tests on various materials

- examines firearms, identifies make, model, caliber; performs examinations to determine if firearm is functional
- examines firearm related evidence—bullets, cartridge cases; compares to test-fired standards
- performs microscopic examinations using a comparison microscope
- performs examinations and comparison of toolmark evidence
- prepares a variety of chemical reagents and solutions using standard forensic laboratory practices
- conducts test firing of weapons to determine penetration, angle, and distance fired
- operates, calibrates, troubleshoots, and performs minor repairs on laboratory instruments including, but not limited to, comparison microscopes, photographic cameras, and video equipment
- obtains latent prints from firearms and toolmark evidence
- performs chemical and microscopic testing on clothing to determine gunshot residue
- performs examinations and comparisons of footwear and tire tread evidence
- prepares scientific reports, interprets and analyzes test results, prepares affidavits, and collects statistical data
- projects a professional image while testifying as an expert in relation to scientific principles and the interpretation of results of examinations
- trains law enforcement personnel, and provides consultation regarding investigative leads and analytical results

Collateral Functions

- stays current in new trends and innovations in the field of forensic firearms and toolmarks, conducts research, and evaluates new procedures and methods, recommending new procedures and updating existing ones

- provides scientific assistance to attorneys, state law-makers, and regulatory agencies

Qualifications

Required Knowledge

- theory and analytical procedures of the natural and physical sciences, including chemistry, organic chemistry, physical chemistry, physics, biology, and allied fields and subdisciplines
- laboratory testing procedures and methods
- equipment and supplies used in a chemistry laboratory
- proper procedures and standard laboratory safety rules and precautions regarding firearms and chemical use
- operational characteristics of all types of firearms, including bullet construction and rifling
- operation and care of comparison microscope, photographic and video equipment
- laboratory evidence collection and preservation procedures
- theory and principles of forensic science
- federal, state, and local laws, codes, and regulations pertaining to forensic science and firearms
- policies, procedures, rules, and regulations governing a forensic firearms laboratory
- operational methods and techniques of forensic laboratory instruments and equipment

Required Skills

- operate, troubleshoot, and perform minor repairs on laboratory instruments and equipment, including comparison microscopes, photographic equipment, and firearms
- work independently with no direct supervision

- review and evaluate new/existing information and analytical techniques for possible implementation into laboratory protocol
- recognize anomalies, formulate hypotheses, and take appropriate action
- provide training to personnel, law enforcement, and others as needed
- work with chemicals and hazards in a safe, noncontaminating manner
- compare and evaluate latent prints from firearms evidence
- communicate clearly and concisely, both orally and in writing
- establish and maintain cooperative working relationships with those contacted in the course of work
- maintain mental capacity that allows the capability of making sound decisions; demonstrate intellectual capabilities
- maintain physical condition appropriate to the performance of assigned duties and responsibilities that may include the following: walking, standing, or sitting for extended periods of time; operating computers and related office equipment
- maintain effective audiovisual discrimination and perception needed for making observations; communicating with others; reading and writing; operating computers, copiers, and related office equipment

Experience and Training Requirements

- Experience: three years of responsible research and practical experience working in a forensic laboratory as a professional firearm/toolmark examiner
- Training: equivalent to a bachelor's degree from an accredited college or university with major course work in criminalistics, forensic science, chemistry, biology, or a related field

Working Conditions

- Environmental conditions: laboratory environments with occasional field visits; exposure to hazardous chemicals
- Physical conditions: essential and marginal functions may require maintaining physical condition necessary for light lifting, sitting and standing for prolonged periods of time

Forensic Laboratory Technician

Provides technical support in a forensic laboratory and maintains evidence control; provides responsible staff assistance to professional laboratory staff; completes a variety of laboratory tasks and procedures as assigned by the laboratory director.

Major Tasks

- collects, logs, and identifies evidence received by the laboratory; assigns accession numbers; maintains security and proper records and storage of evidence in the database
- performs a variety of basic analytical tests and examinations; reviews results with professional laboratory staff; documents results; calibrates lab equipment
- prepares standard chemical solutions and reagents to specific requirements
- test fires weapons, performs microscopic scanning of cartridge cases for Drugfire entry
- prepares gels, overlays, and buffers, including adjusting proper pH of solutions
- maintains computerized log of chemicals, antisera, buffers, and reagents used by lot number, date received, and date opened
- prepares laboratory kits; cleans and disinfects laboratory equipment and specialized glassware
- operates computer terminal

Collateral Functions

- prepares and maintains laboratory inventory

Qualifications

Required Knowledge

- basic principles of organic and inorganic chemistry
- complex mathematical principles
- basic laboratory testing procedures and techniques
- recordkeeping methods and techniques
- equipment and supplies used in a chemistry laboratory
- proper procedures and standard safety precautions in a laboratory environment
- laboratory terminology and techniques, including safety
- use of analytical balances, pH meters, basic microscopes, and computers
- firearms and ammunition, including safe handling
- data entry codes used in the laboratory

Required Skills

- learn to use and understand the operational methods and techniques of laboratory materials and equipment
- learn evidence collection and records maintenance techniques
- learn federal, state, and local laws, codes, and regulations pertaining to a forensic laboratory
- learn policies, procedures, rules, and regulations governing a forensic laboratory
- maintain accurate and complete records
- communicate clearly and concisely, both orally and in writing
- establish and maintain effective working relationships with those contacted in the course of work

- type at a speed necessary for successful work performance
- maintain effective audiovisual discrimination and perception needed for making observations, communicating with others, reading and writing, operating assigned equipment
- maintain mental capacity that allows the capability of making sound decisions; demonstrate intellectual capabilities

Experience and Training Requirements

- Experience: one year of practical experience working in a chemistry laboratory is desirable
- Training: equivalent to an associate's degree or two years of technical or college-level course work in the basic sciences, including successful completion of chemistry/ laboratory classes

Licensing and/or Certification

- possession of, or ability to obtain, an appropriate, valid driver's license

Working Conditions

- Environmental conditions: laboratory environment; exposure to potentially hazardous chemicals; handle evidence of all types, some of which may be hazardous
- Physical conditions: essential and marginal functions may require maintaining physical condition necessary for light lifting and standing for prolonged periods of time

Latent Print Examiner

Conducts fingerprint comparisons of latent prints and finger and palm print exemplar files; performs various tasks relative to assigned areas of responsibility.

LPE I: The entry/trainee-level class of the Latent Print Examiner series. Performs the full range of duties assigned, including the examination and identification of fingerprints. Employees at this level receive only occasional instruction or assistance as new or unusual situations arise, and are fully aware of the operating procedures and policies of the work unit.

LPE II: The full journey-level class in the Latent Print Examiner series. Distinguished from the LPE I level by the degree of proficiency required in comparing prints and the complexity of duties assigned. Employees perform the most difficult and responsible types of duties assigned to levels within the LPE series, including training lower-level employees, and may provide technical supervision to LPE Is. Positions in this class are flexibly staffed and are normally filled by advancement from the LPE I level; when they are filled from outside, applicants have the requisite experience, training, and certification requirements.

Major Tasks (I/II)

- conducts fingerprint comparisons of latent prints and fingerprint exemplars; establishes identity and nonidentity
- processes items of evidence for latent prints; evaluates, searches, and conducts examinations; determines chemical priority; and recovers latent fingerprints photographically or conventionally
- encodes latent prints into the Automated Fingerprint Identification System (AFIS); ensures use of proper methods of coding
- utilizes special photographic techniques in examining prints; determines evidentiary value
- prepares court exhibits; provides expert testimony on latent print examinations
- accepts, logs, and secures physical evidence; safely packages evidence; and ensures integrity of evidence

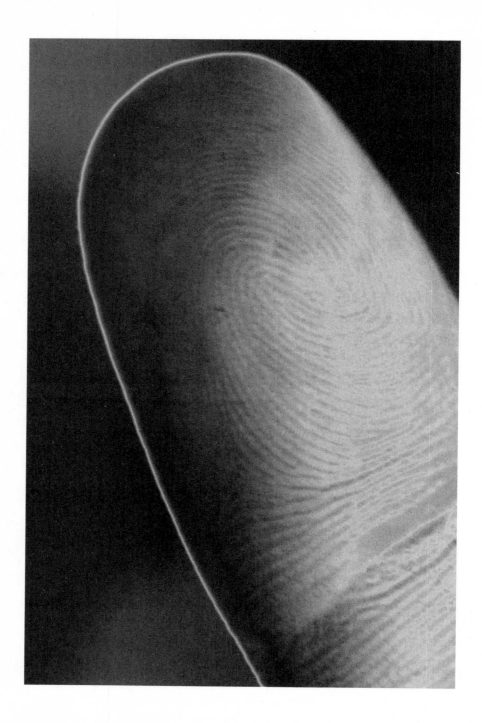

- provides training in proper fingerprinting techniques to resident patrol, jail officers, and investigators
- prepares a variety of reports regarding latent print examinations; maintains statistical data on latent fingerprint examinations
- trains and provides technical supervision to Latent Print Examiner Is

Collateral Functions

- maintains a variety of latent prints, fingerprints, and court records and files
- stays abreast of new trends and innovations in the field of latent print examinations

Qualifications

Required Knowledge (LPE I)

- operations, services, and activities of latent print examinations
- principles and practices of fingerprint science
- modern and complex methods and techniques of latent print examination, identification, and interpretation
- Henry and NCIC classification systems
- modern and complex methods and techniques of latent print processing using powders, chemicals, and Polylite
- departmental policies and procedures
- statutes pertaining to latent print and fingerprint examination
- pertinent federal, state, and local laws, codes, and regulations
- proper safety precautions

Required Knowledge *(LPE II)*

- all knowledge of Latent Print Examiner I
- evidence collection and preservation techniques
- theory and principles of forensic science

Required Skills (LPE I)

- examine latent fingerprints and interpret results
- use and operate photographic and laboratory equipment, including the Polylite
- operate the AFIS and SCOPE computer systems
- interpret and explain department policies and procedures
- prepare clear and concise reports
- communicate clearly and concisely, both orally and in writing
- work efficiently in high-stress environments
- establish and maintain effective working relationships with those contacted in the course of working, including department officials and the general public
- maintain physical condition appropriate to the performance of assigned duties and responsibilities, which may include the following: walking, standing, or sitting for extended periods of time; operating assigned equipment
- maintain effective audiovisual discrimination and perception needed for making observations; communicating with others; reading and writing; operating assigned equipment and vehicles
- maintain mental capacity that allows the capability of making sound decisions; demonstrate intellectual capabilities

Required Skills *(LPE II)*

- LPE I abilities

- ability to work independently in the absence of supervision
- provide training to Latent Print Examiner Is

Experience and Training Requirements

- Experience: LPE I; three years of supervised full-time experience as a Latent Print Examiner. LPE II; five years of supervised full-time experience as a Latent Print Examiner
- Training: LPE I/II; equivalent to a bachelor's degree from an accredited college or university with major course work in criminal justice, forensic science, physical science, or a related field, including specialized training in fingerprint examination

Licensing and/or Certification (I/II)

- possession of, or ability to obtain, an appropriate, valid driver's license
- possession of, or ability to obtain, the International Association for Identification (IAI) Certification

Working Conditions

- Environmental conditions: Office environment; exposure to hazardous chemicals
- Physical conditions: Duties may require maintaining physical conditions necessary for sitting for prolonged periods of time

Photo Technician

Provides photographic support to LVMPD and neighboring police agencies; operates and maintains complex photographic equipment; performs a variety of tasks relative to assigned areas of photography.

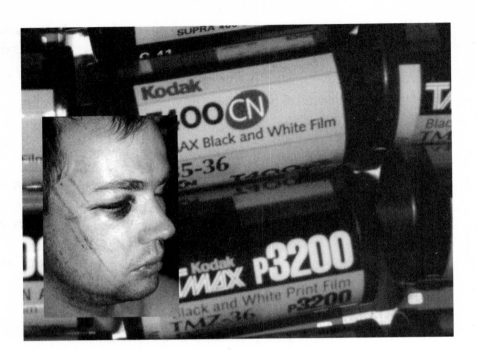

Major Tasks

- processes and prints black and white and color prints, and proofs manually or using photographic processing equipment
- mixes photographic chemicals, monitors and fills chemical tanks, uses proper safety precautions
- cuts, prints, and packages mugs photos and proofs
- pulls negatives and files accordingly
- cross-references negatives to ensure correct numerical identifications; references SCOPE or other data
- takes photographs for police lineups, public relations, and portraits for LVMPD use
- makes minor repairs to and maintains photographic equipment; programs printers, as appropriate

- operates copy camera
- serves as lab receptionist

Collateral Functions

- stays abreast of new trends and innovations in the field of photography
- fills in during supervisor's absence

Qualifications

Required Knowledge

- modern and complex methods and techniques of photographic processing
- principles and practices of color and black and white developing and printing
- chemical mixing procedures
- material and equipment used in a photo lab
- departmental policies and procedures
- pertinent federal, state, and local laws, codes, and regulations

Required Skills

- print and process color and black and white film
- use, operate, and maintain photo processing equipment
- prepare clear and concise reports
- work independently in the absence of supervision
- explain department policies and procedures
- communicate clearly and concisely, both orally and in writing
- establish and maintain effective working relationships with those contacted in the course of work, including department officials and the general public

- maintain physical condition appropriate to the performance of assigned duties and responsibilities, which may include the following: walking, standing, or sitting for extended periods of time; operating assigned equipment
- maintain effective audiovisual discrimination and perception needed for making decisions; discriminating color tones in producing photographs; communicating with others; reading and writing; operating assigned equipment and vehicles
- maintain mental capacity that allows the capability of making sound decisions; demonstrate intellectual capabilities

Experience and Training Requirements

- Experience: two years of responsible photography experience
- Training: equivalent to completion of the twelfth grade supplemented by specialized photography training

Working Conditions

- Environmental conditions: photo laboratory environment; exposure to chemicals
- Physical conditions: duties may require maintaining physical condition necessary for moderate or light lifting, sitting for prolonged period of time
- Reports to Photo Technician Supervisor(s) with supervisory/training/administrative functions

Financial Compensation for Entry-Level LVMPD Civilian Personnel

Evidence Custodian	$31,449.60
Criminalist I/II	$45,545.16
Crime Scene Analyst I	$34,713.12
Document Examiner	$50,273.52
Firearms/Toolmark Examiner	$50,273.52
Forensic Laboratory Technician	$32,235.84
Latent Print Examiner II	$45,545.76
Photo Technician	$25,182.48

Worth it?

You'll have to be the judge. But a decent cab driver in the same city earns about $42,000 annually.

"Help Wanted" in the Real World!

Chemist

The U.S. Army Criminal Investigation Laboratory in Ft. Gillem (GA) is seeking applicants for the position of Chemist. Qualifications include graduation from an accredited four-year college or university with a degree in physical science, life science, or engineering, with one year of specialized experience or one year of graduate-level education or superior academic achievement. Responsibilities include full performance employee services as a chemist assigned to the

Trace Evidence Devision of USACIL. Incumbent will perform developmental assignment under supervision until full performance level is reached, interpret and evaluate the results of examinations performed, and testify as an expert witness. Worldwide travel is required. Salary: $29,539–$81,009.

Criminalist I

The Miami-Dade Police Department Crime Lab is seeking applicants to fill four positions as Criminalist I. Qualifications include a B.S. degree in biology, biochemistry, chemistry, forensic science, or a related field. Primary interest in DNA examiners. Other positions may become available soon. Responsibilities include conducting preliminary examination of physical evidence using serology and DNA analysis technology (STR), reporting findings, and testifying in court. Applicant must successfully pass background, fingerprint, polygraph, and physical examination prior to being hired. Salary: $34,500.

Latent Print Examiner

The Utah Bureau of Forensic Services, State Crime Lab, is seeking applicants for the position of Latent Print Examiner I. Qualifications include a bachelor's degree in criminal justice (augmented with chemistry course work), criminalistics, biochemistry, chemistry, forensic science, or a related field, together with training and/or experience in fingerprint examination. Must be physically able to respond to crime scenes in hazardous environments, including methamphetamine labs, wear a respirator and Tyvec suit, lift and carry 30 to 40 pounds, crawl and work in confined areas, and work all hours (on call). Salary: $27,164–$33,758.

Serologist/DNA Analyst

The Montgomery County Crime Laboratory is seeking applicants for the position of Serologist/DNA Analyst. Qualifications include graduation from an accredited college or university with a bachelor's degree in biology, chemistry, forensic science, or a related field and three years experience in DNA analysis. Preference will be given to those who meet DAB education guidelines, or have experience using the ABI 310, or have basic serology experience. Responsibilities include examining evidence for biological fluids; analyzing blood, semen, and saliva; analyzing samples using STR DNA technology; interpreting test results; maintaining proper QA/QC; performing validation studies; writing reports; testifying in court; using the CODIS database; assisting investigators at crime scenes; and training investigators. Salary: $42,478–$70,411.

Forensic Specialist I/II

The City of Concord (CA) is seeking applicants for the position of Forensic Specialist I/II. Qualifications include a high school diploma or equivalent. For the Level I position, college-level work in forensic science or crime scene investigations is desired. For the Level II position, an associate's degree or two years of complete college course work in criminology, forensic science, or the life sciences is desired. Must have one year (Level I) or two years (Level II) of experience in a law enforcement agency, with duties in the technical tasks associated with the identification, collection, and preservation of evidence, or as a fingerprint examiner. A forensic or scientific background is desirable. Responsibilities include performing crime scene investigations using various technical equipment for the collection, preservation, and documentation of evidence; developing, securing, and packaging physical evidence for the scientific evaluation and comparison of evidence; preparing detailed reports on the observations and activities at crime scenes; testifying in court regarding the findings and processing methods used at a scene, as well as testifying as an expert witness in the presentation of courtroom evidence. The Forensic Specialist must be able to apply scientific expertise in crime scene evidence identification and processing with the ultimate goal of providing accurate information to be used in courts of law. The Forensic Specialist will be proficient in the use of laser lighting systems, chemical processing, photography (35mm cameras), video cameras, and digital cameras. Salary: $38,868–47,256 (Level I); $45,240–$54,984 (Level II).

Crime Scene Technician

The Coral Gables (FL) Police Department is seeking applicants for the position of Crime Scene Technician. Minimum qualifications include graduation from high school or GED, completion of specialized training programs related to forensic science and criminal investigations, and one year of experience in crime scene processing, forensic work, or related field. A comparable amount of training, education, or experience may substitute for minimum qualifications. Must possess a valid Florida driver's license. Must also pass a written test for Crime Scene Technician and obtain certification by the International Association for Identification within 18 months of employment. Responsibilities include performing specialized, technical work gathering evidence at crime scenes; detecting, collecting, preserving, packaging, and transporting evidence; processing latent fingerprints; performing forensic photography and producing crime scene drawings; preparing comprehensive written reports; testifying as an expert witness in court; and operating vans, trucks, hand and power tools, laboratory and camera equipment. Duties involve strenuous physical activity under severe working conditions. The City does not employ individuals who now use or have used tobacco products within the last twelve months. Salary: $31,369–$42,038.

Education Tracks to Positions in Crime Scene Investigations

Just as positions, job titles, and duties in crime scene investigations change with the region and jurisdiction in which they operate, the roads leading applicants to those positions vary widely: postdoctoral degrees, specialist two-year technical programs emphasizing hands-on work, traditional graduate degree programs in a science field, full-fledged forensic programs preparing students for specialization in DNA studies or the chemistry of bombs, even at-home study programs completed by mail. Some people, like anthropologist Dwayne Barbie, even fall into crime scene investigation completely by accident.

"I wasn't in the least interested in crime, at least not modern crimes." Barbie's shelves are full of bones that, found anywhere other than an anthropologist's office, could attract serious attention from law enforcement. "I could tell you the approximate incidence of scurvy in the crew of a sunken British ship of the early eighteenth century or how old an Egyptian prince was when he died, but until I got called out to help dig up a cat, I'd never even thought of the criminal applications of anthropology."

A cat?

"I was visiting family in a rural area for the holidays. One of the locals called the police station to report a skull found in a compost heap. My sister-in-law, who was the station's communications officer, couldn't get hold of anyone at the state lab but, figuring bones were bones, she volunteered my services. I missed Thanksgiving dinner and spent half the afternoon excavating a pile of rotten kitchen refuse, all to uncover a cat's skeleton." Sixteen years later, he's still shaking his head over it. "They were afraid it was an infant. Didn't take more than a glance to assure them it wasn't, but somehow, in the time it took to get that skull out of the garbage, I was hooked."

Many academic specialists, first called in as expert witnesses or impromptu field-workers, discover the rush of this demanding detective work through the back door. Xi Ping, who used to spend his days fitting braces on adolescent teeth, peers at a tiny dental X-ray as he shakes blackened lumps out of an evidence envelope and onto a sterile tray, then tugs the cover off a binocular microscope. "These are from a suspected arson. None of the people who should have been in there are missing, but the janitor used to come in to work late if his day job hung him up and we can't seem to find him, so. . . ." He pops the first tooth under the scope. "We're hoping to match something here with his last X-ray." Something comes into focus and he squints at the X-ray again. "The first time I went to court it was simply luck of the draw. I'd done some bridgework on a suspect in a rape case. I testified that the X-rays done at the time were taken by me on such-and-such a date, that was it. But watching what the real expert could do, matching those X-rays and a bite mark—it was amazing!" The blackened piece gets a brisk rub

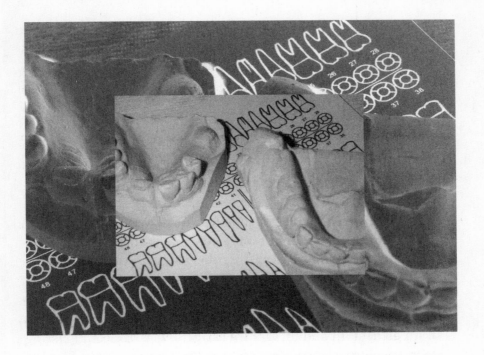

and he peers into the lenses again. "I took a year's sabbatical, went back to school, took some forensic work, and haven't looked back."

Not everyone starts off with a specialist background. In fact, up until recently, there were no schools or jobs for dedicated criminalists except that of detective or print expert. As a result, many jurisdictions, like Dade County in Florida and Las Vegas, recruit laypeople and train them in-house. The apprenticeship ideal, where new workers learn from their experienced superiors, is still very much in evidence. But as more schools open their doors, even entry-level applicants are expected to gain some theoretical and practical experience before beginning their careers.

Associate degrees (generally two-year diploma or certificate programs) cover the basics of crime scene work and the more routine procedures required. As an example, this is how the Forensic Technology Program at Grossmount College in California describes the goals of its students and curriculum: "To provide a broad base of experience, to prepare students who want to work as crime scene technicians. Criminalists tend to specialize, so we give them a taste and appreciation of the various specialties so they can work in the field and return later for more advanced studies like forensic chemistry or forensic biology when they make an informed decision about which areas they want to pursue."

Even at the forensic technology level, however, specialization is already taking place with the program dividing into evidence technology and fingerprint tracks about midway through. "All students learn the basics of both areas, but their hands-on components and theory classes are divided for more advanced techniques and study of the non-routine situations possible."

Some two-year programs, especially those aimed at other law enforcement personnel and related fields (insurance investigators, medical staff, and private investigators) assume an existing familiarity with the technology involved and the concepts of crime scene preservation and chain of custody. Generally, these programs (which are designed to accommodate working schedules) are broken down into specific components that focus on one topic or issue. Courses can be taken independently of one another and combined toward either basic or advanced certificates.

Fullerton University offers its Certificate in Applied Forensic Science through the Extended Education department. Students there are also permitted to complete a Basic Level with four core classes and the completion of three electives, then apply that level's work with seven further electives for an Advanced Level Certificate. The variety of electives (including such apparently tangential subjects as workplace safety) allows for a broad base of student interest. This helps future investigators gain an understanding of fields related to their own where they can find the experts they might need during their own investigations.

A wide array of options open up for those who decide early on

that they hope to find a career in laboratory-based criminalistics or some field of forensic science, and who are prepared to commit to full-time study for the duration of a bachelor's degree.

John Jay College of Criminal Justice in New York, for example, offers a Bachelor of Forensic Science with two tracks: criminalistics or toxicology. The first two years of common study are heavily weighted toward law and the sciences. Physics, biology, and several branches of chemistry require twelve to eighteen hours of lab time and a similar period of theory per semester. Specific courses in forensic science begin in the third year, and biochemistry and physical chemistry are added to the stiff science requirements. In the fourth year, all students complete a ten-week internship in either a criminal or toxicology lab. The program is demanding but presents the graduate with a wider range of career options with higher starting salaries.

Another option at the bachelor's degree–level of study is a more traditional major study area, like biology or physics, but with a decided slant toward immediate applications in forensics. The University of Tennessee at Knoxville, for example, currently offers a Bachelor of Science program with major course work in forensic anthropology. Briefly, this is how the school describes its degree: "In accordance with the aims and goals of forensic anthropology, the focus of the program is the application of skeletal biological techniques to the identification of decomposing and skeletal remains for law enforcement and medicolegal agencies and investigations." Their graduates are anthropologists in every sense of the word, with background in the cultural and archaeological aspects of the work, but with an emphasis on and an understanding of how anthropology principles can work in the field of forensics.

Perhaps one of the most commonly presented aspects of forensic anthropology is reconstruction—that near-magical ability to put a recognizable face on a skull. While practitioners are certainly artists in their medium, reconstruction is by no means guesswork. Anthropologists gain a unique understanding of what the outside looks like by knowing the inside, and a typical course of studies includes comparative anatomy, mammology, and gross human anatomy, courses that sound like they belong in the school of medicine. Add in the technician's side of the equation—electron microscopy, zooarcheology, as well as dozens of science courses—and it becomes rapidly apparent that this art is dependent on pretty serious science.

Further along the spectrum of forensic science studies are the graduate programs. Students in these courses already have a bachelor's or master's degree in a related area. TV criminalist Gil Grissom and real-life forensic scientist Deloit Purgess share a background in entomology, an expertise in insects. "My original career choice, agriculture, bored me to tears in the first year. There's only so many ways to tell a farmer he's got potato blight. Government work gave me more to do, but just wasn't lab-oriented enough. And I'm a real beaker-head. Love to start with a problem and dig until I've got an answer. Perfect personality type for crime work—I was even used to all the paperwork."

That wasn't the only adjustment Purgess needed to make, however, and he was happy enough to get into a master's program in forensic science. "Forensics is so multidisciplinary, and there's so many things important in an investigation that you just don't think of in other fields. Any lab crew knows the importance of tight controls, avoiding contamination, and documenting everything eight different ways. That wasn't what I learned in the master's program. I got to see the criminal justice system from the view of the end-user, the courts, and to see what, beyond the obvious, crime labs can do to make the system work better."

The University of New Haven offers a program very similar to Purgess's, a master's degree in forensic science, and entry standards are high. Applicants must have a B average in undergrad science and math courses, an overall 3.0–4.0 grade average, and letters of recommendation from people in the field, and

they must show high levels of performance and potential during in-person interviews. Once admitted, a stiff slate of studies awaits. Crime scene investigation, laboratory investigation (I/II), criminal procedure, criminal law, law and evidence, physical analysis in forensic laboratory science, biomedical methods, and, depending on concentration, medicolegal investigation and identification, forensic toxicology, analytic chemistry, drug chemistry and identification, forensic serology (I/II), and forensic microscopy are basic to the criminalistics track. However, as forensic science implies interdisciplinary study by its very nature, the remainder of the repertoire of a typical student could include borrowings from other tracks, like the fire science option, which includes elements of structural architecture, fire scene reconstruction, forensic studies of burn victims, and dozens more theory classes on building codes, the chemistry of arson crimes, and the psychology of arsonists. Additionally, a thesis, an original and independent research project, and laboratory survey courses are all required before graduation. Not surprisingly, most students take two to three years to complete the program.

As forensic science programs become more prevalent, they're also becoming more standardized, are more pointedly addressing the role of the criminalist or forensic scientist in the courtroom, and are ensuring that crime scenes and evidence are well documented. Virginia Commonwealth University recently revamped its forensic-related courses, amalgamating them into a cohesive group of studies, which includes English courses devoted to writing concisely and clearly in report form and computer science with references to major criminal justice databases and interfaces, as well as laboratory and theoretical studies.

A Sample Program

The criminalist and forensic scientist of the future is becoming less and less likely to enter the field with no previous experience. Instead, programs like the one at VCU are becoming the standard.

Virginia Commonwealth University
Forensic Science Core Courses (44 credits)

- Intro to Biological Science I/II
- Intro to Biological Science I/II Laboratory Session
- Biotechniques Laboratory Session
- Intro to Molecular Biology
- General Chemistry I/II
- General Chemistry I/II Laboratory Session
- Organic Chemistry I/II
- Organic Chemistry I/II Laboratory Session
- Criminalistics and Crime Analysis
- Criminalistics and Crime Analysis Laboratory Session
- Forensic Evidence and Criminal Procedure
- Forensic Science I/II
- Crime Scene Search and Recovery
- Computer Concepts and Applications
- Business and Technical Report Writing
- Writing for the Workplace
- Intro to Life Sciences
- Precalculus Mathematics
- Practice of Statistics
- General Physics
- Additional course work drawn from Organic Chemistry, Physical Chemistry, Quantitative Analysis, Methods of Organic Synthesis, Calculus, Analytic Geometry, Biochemistry, Forensic Molecular Biology, Population Genetics
- Further course work in Written and Oral Communications, Ethical Principles, two credits in the Visual and Performing Arts (to become aware of body language and its potential, voice control, etc.), Human Behavior, and Urban Environmental Studies

- Specialist courses from Medicinal Chemistry, Clinical Immunology, Justice System Survey, Criminal Law, Foundations of Criminal Justice, Forensic Anthropology, Principles of Death Investigations, Forensic Ecology, General Physics, and Forensic Entomology
- Courses in Education Methods and General Psychology are strongly recommended
- Internship program

This four-year program leaves little room for electives like basket-weaving—though it's likely that some criminalist somewhere would eventually find *that* useful as well!

It would seem that no area of study wouldn't, in some way, benefit the criminalist, which explains the one trait seemingly essential for forensics—curiosity.

"I learn something new every day," says Bill Bonner. "Either in the field, or from colleagues who've stumbled over a totally new situation or new application to an old situation. Or through desperation just trying something to see if it will work. A convention of criminalists, arson investigators, medical examiners, or any combination thereof is an event in itself. If you're not going to keep up with the literature in your field—and several other fields besides—you're quickly left behind. Anyone going into forensics had better be a natural student, someone who enjoys the challenge and—if they're serious about the work—better not be frustrated by the time it can take to implement practical uses of new technology. No field is stagnant. We've been lifting fingerprints forensically for a hundred years, and we're still learning."

And it's not just the science and methodology. Laws, codes, and policies change all the time—to tighten up quality control in the lab, to ensure smoother operation in the field, to help law enforcement personnel communicate with each other and with the laypeople in a jury pool.

On Bonner's desk right now are a journal article on victomol-ogy, a copy of the suggested amendments to the state's policy on issuing search warrants, two large bottles of a new fingerprint powder that might show some promise, and an invitation to speak to a group of convicts at the medium-security prison up the road.

He's still waffling on the invitation.

"Fact is, the day I go into work and don't see something new, I'll know I'm dead and my body just hasn't figured it out yet."

Where does he advise that people start a course of study?

"Anywhere. Anywhere with an accredited program, at least. But I know an awful lot of excellent crime scene people, crimi-nalists, and specialists who started with the major home study programs."

Crime scene investigations by mail?

"Not exactly. It's the theory behind the processes, the basic knowledge everyone in the field should have. Things like whether you powder or fume with Super Glue first when look-ing for prints on different surfaces, or how to recover particular evidence if you can't move it. A lot of authorities require in-coming personnel to have some grasp of the lingo, and courses like those needed for the American Institute of Applied Sci-ence's Forensic Science Certificate certainly give you that. It's actually a requirement for a lot of entry-level positions in many jurisdictions. The LVMPD in Nevada, just as an example, re-quires all its crime scene analysts to obtain the AIAS Forensic Science Certificate, so it's not just a case of dummy courses—they have a lot to teach. That course is part of the program at the USACL (United States Army Crime Laboratory), and a lot of print people start there and are certified by the International As-sociation for Identification."

Why not start with a program like Virginia Commonwealth's?

"VCU has an excellent school of forensic science, and that's a wonderful place to go if you're sure this is for you. AIAS's pro-gram lets you test the waters before you give up your day job.

The tuition is reasonable, the basics get covered well, and you can work it in around your other employment until you want to make the leap."

For those already involved in law enforcement, there's one more option available. It's provided by agencies like the Federal Bureau of Investigation through its various "police schools." Through mobile classrooms and at its training center in Quantico, Virginia, the FBI hosts dozens of classes and seminars at no cost. They are open to all levels of law enforcement. Most interesting to forensic personnel is the Forensic Science Research and Training Center (FSRTC), which regularly sponsors the Honors Intern Program for students who are already enrolled in a science degree program and have an interest in criminalistics.

American Institute of Applied Science
Course Work, Forensic Science Certificate

Basic Forensic Science Program

- 101F Fingerprint Classification and Identification
- 101M Modus Operandi
- 101C Criminal Investigations
- 101FI Firearms Identification
- 101Q Questioned Documents
- 101P Police Photography

Course cost: $808 if course components taken as a unit.

Advanced Forensic Science Program

- 201F Fundamentals of Forensic Investigation
- 201T Trace Evidence and Its Significance
- 201A Fundamentals of Arson and Explosion Investigation
- 201B The Significance of Blood in Criminal Investigations
- 201D Forensic Investigation into Drugs and Alcohol
- 201V Document and Voice Identification

- 201FT Firearms, Toolmarks, and Footwear Impressions
- 201I Investigating Forensic Science on the Internet

Course Cost: $478 if components taken as a unit.

Typical Course Content
101F Fingerprint Classification and Identification
(26 Lessons)

This course covers the methodology of locating latent prints and matching latent prints to known prints, and the classification of prints. Modern technology regarding latent print development, alternative light sources, and the future of computerized classification are also explored.

1. The History of Fingerprints
2. Ridge Formation and Ridge Destruction
3. Pattern Types and Pattern Interpretation
4. Pattern Interpretation II
5. Ridge Characteristics, Ridge Counting, Ridge Tracing
6. The Fingerprint Kit, Recording Fingerprints
7. The Fingerprint Card, Unusual Circumstances
8. Latent Prints, Crime Scene Procedures
9. Latent Prints II
10. Primary Classification
11. Unlettered Loop Subclassification
12. Whorl Subclassification
13. Lettered Loop Subclassification
14. Review—Pattern Interpretation and the Henry Classification

{ "Every murderer is probably someone's
old friend."
AGATHA CHRISTIE }

A SELECTION OF INSTITUTIONS OFFERING PROGRAMS WITH FORENSIC INTEREST

Albany State University
Criminal Justice Department/Forensic Science
504 College Drive
Albany, GA 31705
912-430-4864

Appalachian State University
A.R. Smith Department of Chemistry
525 River Street
Boone, NC 28608
828-262-3010

Baylor University
P.O. Box 97326
Waco, TX 76798–7326
254-752-9284

California State University, Fullerton
2600 E. Nutwood Avenue, Suite 770
Fullerton, CA 92831–3112
714-278-2611

California State University, Los Angeles
Department of Criminal Justice
5151 State University Drive
Los Angeles, CA 90032–8163
801-343-4610

California State University, Sacramento
School of Health and Human Services
Division of Criminal Justice
6000 J Street
Sacramento, CA 95819–6085
916-278-6487

Eastern Kentucky University
Forensic Science Program
521 Lancaster Avenue
Richmond, KY 40475–3102
606-622-1456

Fitchburg State College
Division of Graduate and Continuing Education
Forensic Nursing
160 Pearl Street
Fitchburg, MA 01420
987-345-2151

Florida International University
Department of Chemistry
Miami, FL 33199
305-348-6211

George Washington University
Columbian School of Arts and Sciences
Department of Forensic Science
2036 H Street, NW
Washington, DC 20052
202-994-6211

Grossmont Community College
8800 Grossmont College Drive
El Cajon, CA 92020
619-644-7323

Jacksonville State University
College of Criminal Justice
Jacksonville, AL 36265–9982
256-782-5733

John Jay College of Criminal Justice,
City University of New York
Forensic Science Graduate Admissions
445 West 59th Street, Room 4205N
New York, NY 10019
212-237-8870, 212-237-8899

Louisiana State University, Baton Rouge
Department of Geography and Anthropology
Anthropology/Forensic Anthropology
Graduate Admissions Office
114 David Boyd Hall
Baton Rouge, LA 70803
225-388-1641

Louisiana State University, New Orleans
1100 Florida Avenue
P.O. Box 142
New Orleans, LA 70119–2799
504-619-8584

Loyola University—New Orleans
Department of Chemistry
6363 St. Charles Avenue
New Orleans, LA 70118
504-865-3240

Marshall University School of Medicine
Office of Research and Graduate Education
1542 Spring Valley Drive
Huntington, WV 25755–9310
304-696-7394

Forensic Science
1401 Forensic Science Drive
Huntington, WV 25701–3628
304-690-4361

Metropolitan State College
Department of Chemistry
P.O. Box 173362
Campus Box 52
Denver, CO 80217–3362
303-556-2610

Michigan State University
School of Criminal Justice
560 Baker Hall
East Lansing, MI 48824–1118
517-355-2197, 517-355-5283

National University
Department of Graduate Forensic Sciences
4141 Camino del Rio South
San Diego, CA 92108
619-563-7427

Forensic Science Program
11255 N. Torrey Pines Road
La Jolla, CA 92037
619-642-8419

Nebraska Wesleyan University
Psychology Department
5000 Saint Paul Avenue
Lincoln, NE 68504–2794
402-465-2430

Northwestern University
Health Science Building
240 East Huron
Chicago, IL 60611
312-266-5965

Ohio State University
College of Biological Sciences, Molecular Genetics
984 Biological Sciences Building
484 West 12th Avenue
Columbus, OH 43210–1292
614-292-8084

Ohio University
Department of Chemistry
Clippinger Laboratories
Athens, OH 45701
740-593-1731

Rio Hondo College
3600 Workman Mill Road
Whittier, CA 90601
562-692-0921

St. John's University
St. Vincent's College
8000 Utopia Parkway
Bent Room 268
Jamaica, NY 11439
718-990-6111

Southeast Missouri State University
School of Graduate Studies
Forensic Science
1 University Plaza
Cape Girardeau, MO 63701
573-651-2221

University of Alabama at Birmingham
Department of Justice Science
University Station
901 South 15th Street
Birmingham, AL 35294–2060
205-934-2069

The Graduate School
Forensic Science
1400 University Boulevard, Suite 511
Birmingham, AL 35294–1150
205-934-8227

University of California, Berkeley
140 Warren Hall
Berkeley, CA 94720
510-642-4587

University of Central Florida
Department of Chemistry
P.O. Box 162366
Orlando, FL 32816–2366
407-823-6205

University of Central Oklahoma
College of Math and Science
Forensic Sciences
100 N. University Drive
Edmond, OK 73034
405-974-3341

Department of Chemistry
405-974-5467

University of Chaminade, Honolulu
Forensic Sciences Program
3140 Waialae Avenue
Honolulu, HI 96816–1578
808-440-4209

University of Florida
College of Medicine
P.O. Box 100275
Gainesville, FL 32610–0275
352-846-1579

University of Illinois at Chicago
Forensic Science Program MC866
833 South Wood Street
Chicago, IL 60612–2250
312-996-2250

University of Louisville
School of Dentistry
Surgical and Hospital Dentistry
501 South Preston Street
Louisville, KY 40202
502-852-5083

University of Maryland, Baltimore
Forensic Toxicology
621 W. Lombard Street, Room 300
Baltimore, MD 21201
410-706-7131

Department of Pathology
Office of Graduate Program Director
10 South Pine Street, MSTF 7-34B
Baltimore, MD 21201
410-706-6518

University of Mississippi
Department of Chemistry
University, MS 38677
601-232-7301

University of New Haven
Forensic Science
300 Orange Avenue
West Haven, CT 06516
203-932-7000

University of North Texas
Health Science Center at Fort Worth
Graduate School of Biomedical Sciences
3500 Camp Bowie Boulevard
Fort Worth, TX 76107–2699
800-511-4723

The University of Southern Mississippi
Department of Polymer Science
Southern Station
Box 10076
Hattiesburg, MS 39406–0076
601-266-4868

University of Tennessee, Knoxville
Department of Anthropology
College of Arts and Sciences
Forensic Anthropology
Knoxville, TN 37996–0552
423-974-4408

Forensic Anthropology Program
250 South Stadium Hall
Knoxville, TN 37996–0760
423-974-4408

University of Texas
Dental Branch
Houston, TX 77225
713-500-3050

University of Wisconsin, Platteville
Department of Chemistry and Engineering Physics
Platteville, WI 53818
608-342-1651

Vermont College of Norwich University
36 College Street
Montpelier, VT 05602
800-336-6794, 800-828-8500

Virginia Commonwealth University
College of Humanities and Sciences
Forensic Science
P.O. Box 843051
Richmond, VA 23284
812-828-6919

Wayne State University
Department of Mortuary Service
Division of Forensic Science
5439 Woodard
Detroit, MI 48202
313-577-9099

Weber State University
1137 University Circle
Ogden, UT 84407–1137
801-626-6148

West Chester University
School of Business and Public Affairs
Department of Criminal Justice

McKelvie Hall
102 Rosedale Avenue
West Chester, PA 19383
610-436-2630

Department of Chemistry
610-436-2881

West Virginia University
Forensic Identification Program
Allen Hall, Suite 707
P.O. Box 6121
Morgantown, WV 26506
304-293-2453

York College of Pennsylvania
Department of Chemistry
Country Club Road
York, PA 17405–7199
717-815-1543

SELECT BIBLIOGRAPHY AND SUGGESTED READING

General Criminalistics and Forensic Science

Bagerth, Vernon J. *Practical Homicide Investigation: Tactics, Procedures, and Forensic Techniques.* Boca Raton, Fla.: CRC Press, 1992.

Brenner, John C. *Forensic Science Glossary.* Boca Raton, Fla.: CRC Press, 1999.

Camenson, Blythe. *Opportunities in Forensic Science Careers.* New York: VGM Career Books, 2001.

Campbell, Andrea. *Forensic Science.* New York: Chelsea House Publishers, 1999.

Davies, Geoffrey. *Forensic Science.* New York: American Chemical Society, 1986.

DeForest, Peter R., Gaensslen, Robert E., and Lee, Henry C. *Forensic Science: An Introduction to Criminalistics.* Toronto: McGraw-Hill Ryerson, 1997.

Eckert, William G. *Introduction to Forensic Sciences.* Boca Raton, Fla.: CRC Press, 1996.

Eliopulos, Louis N. *Death Investigator's Handbook: A Field Guide to Crime Scene Processing, Forensic Evaluations, and Investigative Techniques.* Boulder, Colo.: Paladin Press, 1993.

Evans, Colin. *The Casebook of Forensic Detection: How Science Solved 100 of the Worlds's Most Baffling Crimes,.* Toronto: John Wiley & Sons, Canada, 1998.

Fisher, Barry A. *Techniques of Crime Scene Investigation.* South Lyon, Mich.: Zipper Books, 1992.

Fridell, Ron. *Solving Crimes: Pioneers of Forensic Science.* New York: Franklin Watts, 2000.

Innes, Brian. *Bodies of Evidence: The Fascinating World of Forensic Science and How It Helped Solve More Than 100 True Crimes.* Pleasantville, N.Y.: Reader's Digest, 2000.

Kinnee, Kevin B. *Practical Investigation Techniques.* Boca Raton, Fla.: CRC Press, 1994.

Mason, J. K. *Forensic Medicine: An Illustrated Reference.* New York: Chapman & Hall, 1993.

Miller, Hugh. *What the Corpse Revealed: Murder and the Science of Forensic Detection.* New York: St. Martin's Press, 1999.

Nickell, Joe and John F. Fischer. *Crime Science: Methods of Forensic Detection.* Lexington, Ky.: University Press of Kentucky, 1998.

Owen, David. *Hidden Evidence: Forty True Crimes and How Forensic Science Helped Solve Them.* Concord, Mass: Firefly Books, 2000.

Robinson, Stephen P. *Principles of Forensic Medicine.* Oxford, England: Oxford University Press, 1996.

Saferstein, Richard. *Forensic Science Handbook, Volume 1.* New York: Prentice Hall, 2001.

Scafuro, Adele C. *The Forensic Stage.* Cambridge, England: Cambridge University Press, 1993.

Scherman, Tony. *Chasing Napoleon: Forensic Portraits.* Glasgow, Scotland: Moffat, Cameron & Hollis, 2000.

Siegel, J. A., Pekka J. Saukko, and Geoffrey C. Knupfer. *Encyclopedia of Forensic Science.* New York: Academic Press, 2000.

Silverstein, Herman. *Threads of Evidence: Using Forensic Science to Solve Crimes.* New York: Twenty-First Century Books, 1996.

Stark, Margaret M. *A Physician's Guide to Clinical Forensic Medicine.* Clifton, N.J.: Humana Press, 2000.

Thomas, Ronald R. *Detective Fiction and the Rise of Forensic Science.* Cambridge, England: Cambridge University Press, 1999.

Ubelaker, Douglas and Scammell, Henry. *Bones: A Forensic Detective's Casebook.* New York: M. Evans and Co., 2000.

Various, Collected. *Handbook of Forensic Science: The Official FBI Guide.* Collingdale, Pa.: DIANE Publishing Company, 1994.

Firearms and Ballistics

Dimaio, Vincent. *Gunshot Wounds: Practical Aspects of Firearms, Ballistics, and Forensic Techniques.* Boca Raton, Fla.: CRC Press, 1998.

Heard, Brian. *Firearms and Ballistics Handbook of Examining and Interpreting Forensic Evidence.* Toronto: John Wiley & Sons, Canada, 1997.

Yinon, Jehuda. *Forensic and Environmental Detection of Explosives.* New York: John Wiley & Sons, 1999.

Impression and Trace Evidence

Bodziak, William J. *Footwear Impression Evidence.* Boca Raton, Fla.: CRC Press, 1992.

———. *Forensic Application of Mass Spectrometry.* Boca Raton, Fla.: CRC Press, 1994.

McDonald, Peter. *Tire Imprint Evidence.* Boca Raton, Fla.: CRC Press, 1992.

Richards, P. *Forensic Microscopy: A Practical Guide.* Berlin, Germany: Springer-Verlag, 2001.

Questioned Document Examination

THE
FORENSIC
CASEBOOK

306

Hilton, Ordway. *Scientific Examination of Questioned Documents.* Boca Raton, Fla.: CRC Publications, 1992.

Hollien, H. *The Acoustics of Crime: The New Science of Forensic Phonetics.* New York: Plenum Publishing, 1990.

Kniffka, Hannes. *Recent Developments in Forensic Linguistics.* New York: Peter Lang Publishing, 1996.

Morris, Ronald N. *Forensic Handwriting Identification.* New York: Academic Press, 2000.

Slyter, Steven A. *Forensic Signature Examination.* Springfield, Ill.: Charles C. Thomas, 1995.

Ridgeology and Fingerprint Science

Beaven, Colin. *Fingerprints: The Origins of Crime Detections and the Murder Case That Launched Forensic Science.* Boston, Mass: Little, Brown, 2001.

Cowger, James F. *Friction Ridge Skin: Comparison and Identification of Fingerprints.* Boca Raton, Fla.: CRC Press, 1992.

Miller, Hugh. *Forensic Fingerprints: Remarkable Real-Life Murder Cases Solved by Forensic Detection.* London, England: Headline, 1998.

Photography and Illustration

Redsicker, David R. *Practical Aspects of Forensic Photography*. Boca Raton, Fla.: CRC Press, 1992.

Russ, John C. *Forensic Uses of Digital Imaging*. Boca Raton, Fla.: CRC Press, 2001.

Taylor, Karen T. *Forensic Art and Illustration*. Boca Raton, Fla.: CRC Press, 1998.

Forensic Pathology, Histology, Toxicology, Serology, DNA Analysis

Alonso, Kenneth and Carmen Alonso. *Forensic Pathology*. Atlanta, Ga.: Allegro Press, 1997.

Committee on Vision Staff National Research Council. *The Evaluation of Forensic DNA Evidence*. Washington, DC: National Academy Press, 1996.

Dix, Jay. *Guide to Forensic Pathology*. Boca Raton, Fla.: CRC Press, 1998.

Evett, Ian W. and Bruce S. Weir. *Interpreting DNA Evidence: Statistical Genetics for Forensic Scientists*. Sunderland, Mass.: Sinauer Associates, 1998.

Farley, Mark A. and James J. Harrington, *Forensic DNA Technology*. London, England: Lewis Publishers, 1990.

Ferner, R. E. *Forensic Pharmacology: Medicines, Mayhem, and Malpractice*. Oxford, England: Oxford University Press, 1996.

Inman, Keith and Norah Rudin. *An Introduction to Forensic DNA Analysis*. Boca Raton, Fla.: CRC Press, 1997.

Jones, Nancy L. *Atlas of Forensic Pathology*. New York: Igaku-Shoin Medical Publishers, 1996.

Knight, Bernard. *Forensic Pathology*. New York: Edward Arnold, 1996.

Lincoln, Patrick J. *Forensic DNA Profiling Protocols*. Clifton, N.J.: Humana Press, 1998.

Loue, Sana. *Forensic Epidemiology: A Comprehensive Guide for Legal and Epidemiology Professionals*. Carbondale, Ill.: Southern Illinois University Press, 1998.

McLay, W. D. *Clinical Forensic Medicine*. Oxford, England: Oxford University Press, 1996.

National Research Council Staff Committee on DNA Technology in Forensic Science. *DNA Technology in Forensic Science.* Washington, DC: National Academy Press, 1992.

Ossulton, M. David. *Forensic Toxicology.* London, England: Taylor & Francis, 1996.

Williams, David J. *Forensic Pathology.* London, England: Churchill Livingstone Press, 1998.

Forensic Computer Science

Casey, Eoghan. *Digital Evidence and Computer Crime: Forensic Science, Computers, and the Internet.* New York: Academic Press, 2000.

Clark, Franklin and Ken Diliberto. *Investigating Computer Crime.* Boca Raton, Fla.: CRC Press, 1996.

Sammes, Jenkinson. *Forensic Computing: A Practitioner's Guide.* Berlin, Germany: Springer-Verlag, 2000.

Thornhill, William T. *Forensic Accounting: How to Investigate Financial Fraud.* Chicago, Ill.: Irwin Professional Publishing, 1994.

Zaenglein, Norbert. *Secret Software: Making the Most of Computer Resources for Data Protection, Information Recovery, Forensic Examination, Crime Investigation and More.* Boulder, Colo.: Paladin Press, 2000.

Animal Examiners

David, Edward, Andrew Rebmann, and Marcella Sorg. *Cadaver Dog Handbook: Forensic Training and Tactics for the Recovery of Human Remains.* Boca Raton, Fla.: CRC Press, 2000.

Forensic Psychology

Ackerman, Marc. *Essentials of Forensic Psychological Assessment.* Toronto: John Wiley & Sons, Canada, 1999.

Hess, Allen K. and Irving B. Weiner. *Handbook of Forensic Psychology.* Toronto: John Wiley & Sons, Canada, 1998.

Kirwin, Barbara. *The Mad, the Bad, and the Innocent: Criminal Mind on Trial—Tales of a Forensic Psychologist.* New York: Harper Collins, 1998.

Miller, Hugh. *Unquiet Minds: The World of Forensic Psychiatry.* Toronto: Musson Books, 1996.

Niehaus, Joe. *Investigative Forensic Hypnosis.* Boca Raton, Fla.: CRC Press, 1998.

Semrau, Stanley and Judy Gale. *Murderous Minds on Trial: Terrible Tales from a Forensic Psychiatrist's Casebook.* Toronto: Dundurn Press, 2001.

Simon, Robert A. *Bad Men Do What Good Men Dream: A Forensic Psychiatrist Illuminates the Darker Side of Human Behavior.* Washington, DC: American Psychiatric Press, 1999.

Forensic Anthropology and Dentistry

Bernstein, Mark L. and James A. Cottone, *Forensic Odontology.* Boca Raton, Fla.: CRC Press, 1999.

Burns, Karen Ramey. *The Forensic Anthropology Training Manual.* New York: Prentice Hall, 1999.

El-Najjar, Mahmoud Y. and K. Richard McWilliams. *Forensic Anthropology: The Structure, Morphology and Variation of Human Bone and Dentition.* Springfield, Ill.: Charles C. Thomas, 1978.

Haglund, William D., Alison Galloway, and Tal Simmons. *Practical Forensic Anthropology of Human Skeletal Remains: Recovery, Analysis, and Resolution.* Boca Raton, Fla.: CRC Press, 1999.

Hunter, John. *Studies in Crime: Introduction to Forensic Archaeology.* London, England: Routledge, 1997.

Jackson, Donna M. *The Bone Detectives: How Forensic Anthropologists Solve Crimes and Uncover Mysteries of the Dead.* Boston, Mass.: Little, Brown, 2001.

Krogman, Wilton M. and M. Yasar Iscan. *The Human Skeleton in Forensic Medicine.* Springfield, Ill.: Charles C. Thomas, 1986.

Maples, William R. and Michael Browning. *Dead Men Do Tell Tales: The Strange and Fascinating Cases of a Forensic Anthropologist.* New York: Doubleday, 1995.

Prag, John and Richard Neave. *Making Faces: Using Forensic and Archaeological Evidence.* Kingsville, Tx.: Texas A & M University-Kingsville Bookstore, 1997.

Rathbun, Ted and Jane Buikstra. *Human Identification: Case Studies in Forensic Anthropology.* Springfield, Ill.: Charles C. Thomas, 1984.

Reichs, Kathleen J. *Forensic Osteology: Advances in the Identification of Human Remains.* Springfield, Ill.: Charles C. Thomas, 1997.

Rhine, Stanley. *Bone Voyage: A Journey in Forensic Anthropology.* Albuquerque, N. Mex.: University of New Mexico, 1998.

Thomas, Peggy. *Talking Bones: The Science of Forensic Anthropology.* New York: Facts on File, 1995.

Bloodstain Evidence

Eckert, William G. and Stuart H. James, *Interpretation of Bloodstain Evidence at Crime Scenes.* Boca Raton, Fla.: CRC Press, 1998.

Forensic Entomology

Byrd, Jason H. and James L. Castner. *Forensic Entomology: Utility of Arthropods in Legal Investigations.* Boca Raton, Fla.: CRC Press, 1999.

Forensics in the Courtroom

Becker, Ronald F. *Scientific Evidence and Expert Testimony Handbook: A Guide for Lawyers, Criminal Investigators, and Forensic Specialists.* Springfield, Ill: Charles C. Thomas, 1997.

Berger, Arthur S. *Dying and Death in Law and Medicine: A Forensic Primer for Health and Legal Professionals.* Westport, Conn.: Greenwood Publishing Group, 1992.

Kantor, A. Tana. *Winning Your Case With Forensic and Demonstrative Graphics.* Boca Raton, Fla.: CRC Press, 1998.

Roberts, Paul and Chris Willmore. *Role of Forensic Science Evidence in Criminal Proceedings.* Lanham, Md.: Bernan Associates, 1993.

Vignaux, G. A. *Interpreting Evidence: Evaluating Forensic Science in the Courtroom.* Toronto: John Wiley & Sons, Canada, 1995.

White, P. C. *Crime Scene to Court: The Essentials of Forensic Science.* Toronto: H. B. Fenn & Co., 1998.

Other Forensic Speciality Applications

Aitken, C. G. *Statistics and the Evaluation of Evidence for Forensic Scientists*. Toronto: John Wiley & Sons, Canada, Limited, 1995.

Bologna, G. Jack and Robert J. Lindquist. *Fraud Auditing and Forensic Accounting: New Tools and Techniques*. Toronto: John Wiley & Sons, Canada, 1995.

Brogdon, B. G. *Forensic Radiology*. Boca Raton, Fla.: CRC Press, 1998.

Carper, Kenneth. *Forensic Engineering*. South Lyon, Mich.: Zipper Books, 2000.

Jungreis, Ervin. *Spot Test Analysis: Clinical, Environmental, Forensic, and Geochemical Applications*. New York: John Wiley & Sons, 1996.

Manning, George A. *Financial Investigation and Forensic Accounting*. Boca Raton, Fla.: CRC Press, 1999.

Mason, J. K. *Pediatric Forensic Medicine and Pathology*. New York: Chapman & Hall, 1989.

Murray, Raymond C. and John C. Tedrow. *Forensic Geology*. Toronto: Prentice-Hall Canada, 1991.

Ratay, Robert T. *Forensic Structural Engineering Handbook*. Toronto: McGraw-Hill Book Company, 2000.

Shuirman, Gerard and James Slosson. *Forensic Engineering: Environmental Case Histories for Civil Engineers and Geologists*. New York: Academic Press, 1992.

INDEX